Big Data Application Architecture Q & A

A Problem-Solution Approach

Nitin Sawant

Himanshu Shah

Apress®

Big Data Application Architecture Q & A

ISBN-13 (pbk): 978-1-4302-6292-3

ISBN-13 (electronic): 978-1-4302-6293-0

President and Publisher: Paul Manning
Lead Editor: Saswata Mishra
Technical Reviewer: Soumendra Mohanty
Editorial Board: Steve Anglin, Mark Beckner, Ewan Buckingham, Gary Cornell, Louise Corrigan, Jim DeWolf, Jonathan Gennick, Jonathan Hassell, Robert Hutchinson, Michelle Lowman, James Markham, Matthew Moodie, Jeff Olson, Jeffrey Pepper, Douglas Pundick, Ben Renow-Clarke, Dominic Shakeshaft, Gwenan Spearing, Matt Wade, Steve Weiss
Coordinating Editor: Mark Powers
Copy Editor: Roger LeBlanc
Compositor: SPi Global
Indexer: SPi Global
Artist: SPi Global
Cover Designer: Anna Ishchenko

Distributed to the book trade worldwide by Springer Science+Business Media New York, 233 Spring Street, 6th Floor, New York, NY 10013. Phone 1-800-SPRINGER, fax (201) 348-4505, e-mail orders-ny@springer-sbm.com or visit www.springeronline.com. Apress Media, LLC is a California LLC and the sole member (owner) is Springer Science+Business Media Finance Inc (SSBM Finance Inc). SSBM Finance Inc is a Delaware corporation.

For information on translations, please e-mail rights@apress.com or visit www.apress.com.

Apress and friends of ED books may be purchased in bulk for academic, corporate, or promotional use. eBook versions and licenses are also available for most titles. For more information, reference our Special Bulk Sales–eBook Licensing web page at www.apress.com/bulk-sales.

Any source code or other supplementary materials referenced by the author in this text is available to readers at www.apress.com/9781430262923. For detailed information about how to locate your book's source code, go to www.apress.com/source-code/.

Contents at a Glance

Contents

About the Authors

Nitin Sawant is managing director, technology, and is the practice lead for technology architecture for BPM, SOA, and cloud at Accenture India. He is an Accenture certified master technology architect (CMTA), leading various initiatives in the emerging technologies of cloud and big data. Nitin has over 17 years of technology experience in developing, designing, and architecting complex enterprise-scale systems based on Java, JEE, SOA, and BPM technologies. He received his master's degree in technology in software engineering from the Institute of System Science, National University of Singapore. He graduated with a bachelor's degree in electronics engineering from Bombay University. He is a certified CISSP, CEH, and IBM-certified SOA solutions architect. Nitin has filed three patents in the SOA-BPM space and is currently pursuing his PHD in BPM security from BITS Pilani, India.

Himanshu Shah is an Accenture senior technology architect with 14 years of IT experience and currently leads the Big Data (Hadoop) Capability in Accenture India Delivery Centre. Himanshu has acted as an enterprise architect for projects involving cloud computing, operations architecture and big data. Himanshu has worked extensively in custom development of JEE based architecture for multiple clients in various Industries including telecom, retail and the insurance domain. Himanshu also has expertise in ITIL operations architecture. Himanshu has been part of Java, Platform Cloud, and SOA Centre of Excellences within Accenture.

About the Technical Reviewer

Soumendra Mohanty has over 17 years of technology experience in developing, designing, implementing and business transformation programs. Soumendra is currently engaged with Mindtree in the role of their global lead for Data and Analytics Services. Soumendra is an industry renowned expert in the BI, analytics and big data arenas. He has been a prolific writer, has published several books and papers and regularly presents in worldwide forums. He received his master's degree in Computers and Applications from College of Engineering and Technology, Orissa University of Agriculture and Technology. Soumendra has filed three patents in the data and analytics space and is currently pursuing his PHD in real time big data and analytics from ITER, India.

Acknowledgments

We wish to acknowledge the support received from our families while we burned the candle on both ends to meet the publishing deadlines. A warm thanks to Soumendra, who is also the reviewer of this book or inspiring us to write this book.

Introduction

Big data is opening up new opportunities for enterprises to extract insight from huge volumes of data in real time and across multiple relational and nonrelational data types. The architectures for realizing these opportunities are based on relatively less expensive and heterogeneous infrastructures than the traditional monolithic and hugely expensive options that exist currently.

The architectures for realizing big data solutions are composed of heterogeneous infrastructures, databases, and visualization and analytics tools. Selecting the right architecture is the key to harnessing the power of big data. However, heterogeneity brings with it multiple options for solving the same problem, as well as the need to evaluate trade-offs and validate the "fitness-for-purpose" of the solution.

There are myriad open source frameworks, databases, Hadoop distributions, and visualization and analytics tools available on the market, each one of them promising to be the best solution. How do you select the best end-to-end architecture to solve your big data problem?

- Most other big data books on the market focus on providing design patterns in the map reduce or Hadoop area only.

- This book covers the end-to-end application architecture required to realize a big data solution covering not only Hadoop, but also analytics and visualization issues.

- Everybody knows the use cases for big data and the stories of Walmart and EBay, but nobody describes the architecture required to realize those use cases.

- If you have a problem statement, you can use the book as a reference catalog to search the corresponding closest big data pattern and quickly use it to start building the application.

- CxOs are being approached by multiple vendors with promises of implementing the perfect big data solution. This book provides a catalog of application architectures used by peers in their industry.

- The current published content about big data architectures is meant for the scientist or the geek. This book attempts to provide a more industry-aligned view for architects.

- This book will provide software architects and solution designers with a ready catalog of big data application architecture patterns that have been distilled from real-life, big data applications in different industries like retail, telecommunication, banking, and insurance. The patterns in this book will provide the architecture foundation required to launch your next big data application.

■ ■ ■

CHAPTER 1

Big Data Introduction

Why Big Data

As you will see, this entire book is in problem-solution format. This chapter discusses topics in big data in a general sense, so it is not as technical as other chapters. The idea is to make sure you have a basic foundation for learning about big data. Other chapters will provide depth of coverage that we hope you will find useful no matter what your background. So let's get started.

Problem

What is the need for big data technology when we have robust, high-performing, relational database management systems (RDBMS)?

Solution

Since the theory of relational databases was postulated in 1980 by Dr. E. F. Codd (known as "Codd's 12 rules") most data has been stored in a structured format, with primary keys, rows, columns, tuples, and foreign keys. Initially, it was just transactional data, but as more and more data accumulated, organizations started analyzing the data in an offline mode using data warehouses and data marts. Data analytics and business intelligence (BI) became the primary drivers for CxOs to make forecasts, define budgets, and determine new market drivers of growth.

This analysis was initially conducted on data within the enterprise. However, as the Internet connected the entire world, data existing outside an organization became a substantial part of daily transactions. Even though things were heating up, organizations were still in control even though the data was getting voluminous with normal querying of transactional data. That data was more or less structured or relational.

Things really started getting complex in terms of the variety and velocity of data with the advent of social networking sites and search engines like Google. Online commerce via sites like Amazon.com also added to this explosion of data. Traditional analysis methods as well as storage of data in central servers were proving inefficient and expensive. Organizations like Google, Facebook, and Amazon built their own custom methods to store, process, and analyze this data by leveraging concepts like map reduce, Hadoop distributed file systems, and NoSQL databases.

The advent of mobile devices and cloud computing has added to the amount and pace of data creation in the world, so much so that 90 percent of the world's total data has been created in the last two years and 70 percent of it by individuals, not enterprises or organizations. By the end of 2013, IDC predicts that just under 4 trillion gigabytes of data will exist on earth. Organizations need to collect this data from social media feeds, images, streaming video, text files, documents, meter data, and so on to innovate, respond immediately to customer needs, and make quick decisions to avoid being annihilated by competition.

However, as I mentioned, the problem of big data is not just about volume. The unstructured nature of the data (variety) and the speed at which it is created by you and me (velocity) is the real challenge of big data.

Aspects of Big Data

Problem

What are the key aspects of a big data system?

Solution

A big data solution must address the three Vs of big data: data velocity, variety, and complexity, in addition to volume.

Velocity of the data is used to define the speed with which different types of data enter the enterprise and are then analyzed.

Variety addresses the unstructured nature of the data in contrast to structured data in weblogs, radio frequency ID (RFID), meter data, stock-ticker data, tweets, images, and video files on the Internet.

For a data solution to be considered as big data, the volume has to be at least in the range of 30–50 terabytes (TBs).

However, large volume alone is not an indicator of a big data problem. A small amount of data could have multiple sources of different types, both structured and unstructured, that would also be classified as a big data problem.

How Big Data Differs from Traditional BI

Problem

Can we use traditional business intelligence (BI) solutions to process big data?

Solution

Traditional BI methodology works on the principle of assembling all the enterprise data in a central server. The data is generally analyzed in an offline mode. The online transaction processing (OLTP) transactional data is transferred to a denormalized environment called as a *data warehouse*. The data is usually structured in an RDBMS with very little unstructured data.

A big data solution, however, is different in all aspects from a traditional BI solution:

- Data is retained in a distributed file system instead of on a central server.

- The processing functions are taken to the data rather than data being taking to the functions.

- Data is of different formats, both structured as well as unstructured.

- Data is both real-time data as well as offline data.

- Technology relies on massively parallel processing (MPP) concepts.

How Big Is the Opportunity?

Problem

What is the potential big data opportunity?

Solution

The amount of data is growing all around us every day, coming from various channels (see Figure 1-1).

As 70 percent of all data is created by individuals who are customers of some enterprise or the other, organizations cannot ignore this important source of feedback from the customer as well as insight into customer behavior.

Content Type	Quantity		Comments
Internet	20	Exabytes (10^{18})	1 exabyte = 1,000,000 terabytes
Web Pages	1.5	Trillion (10^{12})	Plus "dark Web"
Tweets	20	Billion (10^9)	50 million user accounts
Live Posts	2.1	Billion (10^9)	Forums, discussion boards
Social Members	2.1	Billion (10^9)	Memberships — top 115 social sites
Social Content Creators	600	Million (10^6)	People (33% of Internet users)
Facebook Members	500	Million (10^6)	40% of online hours, top 10 properties
YouTube Visitors	375	Million (10^6)	As of December 2009
Blogs	70	Million (10^6)	36,718 listed on Technorati
Formal Periodicals	10s	Thousands (10^3)	Newspapers, other publications

Figure 1-1. *Information explosion*

Big data drove an estimated $28 billion in IT spending last year, according to market researcher Gartner, Inc. That figure will rise to $34 billion in 2013 and $232 billion in IT spending through 2016, Gartner estimates.

The main reason for this growth is the potential Chief Information Officers (CIOs) see in the greater insights and intelligence contained in the huge unstructured data they have been receiving from outside the enterprise. Unstructured data analysis requires new systems of record—for example, NoSQL databases—so that organizations can forecast better and align their strategic plans and initiatives.

Deriving Insight from Data

Problem

What are the different insights and inferences that big data analysis provides in different industries?

Solution

Companies are deriving significant insights by analyzing big data that gives a combined view of both structured and unstructured customer data. They are seeing increased customer satisfaction, loyalty, and revenue. For example:

- Energy companies monitor and combine usage data recorded from smart meters in real time to provide better service to their consumers and improved uptime.

- Web sites and television channels are able to customize their advertisement strategies based on viewer household demographics and program viewing patterns.

- Fraud-detection systems are analyzing behaviors and correlating activities across multiple data sets from social media analysis.

- High-tech companies are using big data infrastructure to analyze application logs to improve troubleshooting, decrease security violations, and perform predictive application maintenance.

- Social media content analysis is being used to assess customer sentiment and improve products, services, and customer interaction.

These are just some of the insights that different enterprises are gaining from their big data applications.

Cloud Enabled Big Data

Problem

How is big data affected by cloud-based virtualized environments?

Solution

The inexpensive option of storage that big data and Hadoop deliver is very well aligned to the "everything as a service" option that cloud-computing offers.

Infrastructure as a Service (IaaS) allows the CIO a "pay as you go" option to handle big data analysis. This virtualized option provides the efficiency needed to process and manage large volumes of structured and unstructured data in a cluster of expensive virtual machines. This distributed environment gives enterprises access to very flexible and elastic resources to analyze structured and unstructured data.

Map reduce works well in a virtualized environment with respect to storage and computing. Also, an enterprise might not have the finances to procure the array of inexpensive machines for its first pilot. Virtualization enables companies to tackle larger problems that have not yet been scoped without a huge upfront investment. It allows companies to scale up as well as scale down to support the variety of big data configurations required for a particular architecture.

Amazon Elastic MapReduce (EMR) is a public cloud option that provides better scaling functionality and performance for MapReduce. Each one of the Map and Reduce tasks needs to be executed discreetly, where the tasks are parallelized and configured to run in a virtual environment. EMR encapsulates the MapReduce engine in a virtual container so that you can split your tasks across a host of virtual machine (VM) instances.

As you can see, cloud computing and virtualization have brought the power of big data to both small and large enterprises.

Structured vs. Unstructured Data

Problem

What are the various data types both within and outside the enterprise that can be analyzed in a big data solution?

Solution

Structured data will continue to be analyzed in an enterprise using structured access methods like Structured Query Language (SQL). However, the big data systems provide tools and structures for analyzing unstructured data.

New sources of data that contribute to the unstructured data are sensors, web logs, human-generated interaction data like click streams, tweets, Facebook chats, mobile text messages, e-mails, and so forth.

RDBMS systems will continue to exist with a predefined schema and table structure. Unstructured data is data stored in different structures and formats, unlike in a a relational database where the data is stored in a fixed row-column like structure. The presence of this hybrid mix of data makes big data analysis complex, as decisions need to be made regarding whether all this data should be first merged and then analyzed or whether only an aggregated view from different sources has to be compared.

We will see different methods in this book for making these decisions based on various functional and nonfunctional priorities.

Analytics in the Big Data World

Problem

How do I analyze unstructured data, now that I do not have SQL-based tools?

Solution

Analyzing unstructured data involves identifying patterns in text, video, images, and other such content. This is different from a conventional search, which brings up the relevant document based on the search string. Text analytics is about searching for repetitive patterns within documents, e-mails, conversations and other data to draw inferences and insights.

Unstructured data is analyzed using methods like natural language processing (NLP), data mining, master data management (MDM), and statistics. Text analytics use NoSQL databases to standardize the structure of the data so that it can be analyzed using query languages like PIG, Hive, and others. The analysis and extraction processes take advantage of techniques that originated in linguistics, statistics, and numerical analysis.

Big Data Challenges

Problem

What are the key big data challenges?

Solution

There are multiple challenges that this great opportunity has thrown at us.

One of the very basic challenges is to understand and prioritize the data from the garbage that is coming into the enterprise. Ninety percent of all the data is noise, and it is a daunting task to classify and filter the knowledge from the noise.

In the search for inexpensive methods of analysis, organizations have to compromise and balance against the confidentiality requirements of the data. The use of cloud computing and virtualization further complicates the decision to host big data solutions outside the enterprise. But using those technologies is a trade-off against the cost of ownership that every organization has to deal with.

Data is piling up so rapidly that it is becoming costlier to archive it. Organizations struggle to determine how long this data has to be retained. This is a tricky question, as some data is useful for making long-term decisions, while other data is not relevant even a few hours after it has been generated and analyzed and insight has been obtained.

With the advent of new technologies and tools required to build big data solutions, availability of skills is a big challenge for CIOs. A higher level of proficiency in the data sciences is required to implement big data solutions today because the tools are not user-friendly yet. They still require computer science graduates to configure and operationalize a big data system.

Defining a Reference Architecture

Problem

Is there a high-level conceptual reference architecture for a big data landscape that's similar to cloud-computing architectures?

Solution

Analogous to the cloud architectures, the big data landscape can be divided into four layers shown vertically in Figure 1-2:

- **Infrastructure as a Service (IaaS):** This includes the storage, servers, and network as the base, inexpensive commodities of the big data stack. This stack can be bare metal or virtual (cloud). The distributed file systems are part of this layer.

- **Platform as a Service (PaaS):** The NoSQL data stores and distributed caches that can be logically queried using query languages form the platform layer of big data. This layer provides the logical model for the raw, unstructured data stored in the files.

- **Data as a Service (DaaS):** The entire array of tools available for integrating with the PaaS layer using search engines, integration adapters, batch programs, and so on is housed in this layer. The APIs available at this layer can be consumed by all endpoint systems in an elastic-computing mode.

- **Big Data Business Functions as a Service (BFaaS):** Specific industries—like health, retail, ecommerce, energy, and banking—can build packaged applications that serve a specific business need and leverage the DaaS layer for cross-cutting data functions.

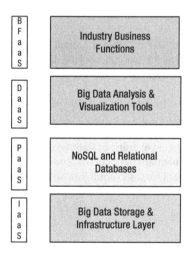

Figure 1-2. *Big data architecture layers*

You will see a detailed big data application architecture in the next chapter that essentially is based on this four-layer reference architecture.

Need for Architecture Patterns

Problem

Why do we need big data architecture patterns?

Solution

Though big data offers many benefits, it is still a complex technology. It faces the challenges of both service-oriented architecture (SOA) and cloud computing combined with infrastructure and network complexities. SOA challenges, like distributed systems design, along with cloud challenges, like hybrid-system synchronization, have to be taken care of in big data solutions.

A big data implementation also has to take care of the "ilities" or nonfunctional requirements like availability, security, scalability, performance, and so forth. Combining all these challenges with the business objectives that have to be achieved, requires an end-to-end application architecture view that defines best practices and guidelines to cope with these issues.

Patterns are not perfect solutions, but in a given context they can be used to create guidelines based on experiences where a particular solution or pattern has worked. Patterns describe both the problem and solution that can be applied repeatedly to similar scenarios.

Summary

You saw how the big data revolution is changing the traditional BI world and the way organizations run their analytics initiatives. The cloud and SOA revolution are the bedrock of this phenomenon, which means that big data faces the same challenges that were faced earlier, along with some new challenges in terms of architecture, skills, and tools. A robust, end-to-end application architecture is required for enterprises to succeed in implementing a big data system. In this journey, if we can help you by showing you some guidelines and best practices we have encountered to solve some common issues, it will make your journey faster and relatively easier. Let's dive deep into the architecture and patterns.

■ ■ ■

CHAPTER 2

Big Data Application Architecture

Enterprises and their customers have become very diverse and complex with the digitalization of business. Managing the information captured from these customers and markets to gain a competitive advantage has become a very expensive proposition when using the traditional data analytics methods, which are based on structured relational databases. This dilemma applies not only to businesses, but to research organizations, governments, and educational institutions that need less expensive computing and storage power to analyze complex scenarios and models involving images, video, and other data, as well as textual data.

There are also new sources of data generated external to the enterprise that CXOs want their data scientists to analyze to find that proverbial "*needle in a haystack.*" New information sources include social media data, click-stream data from web sites, mobile devices, sensors, and other machine-generated data. All these disparate sources of data need to be managed in a consolidated and integrated manner for organizations to get valuable inferences and insights. The data management, storage, and analysis methods have to change to manage this big data and bring value to organizations.

Architecting the Right Big Data Solution

Problem

What are the essential architecture components of a big data solution?

Solution

Prior to jumping on the big data bandwagon you should ensure that all essential architecture components required to analyze all aspects of the big data set are in place. Without this proper setup, you'll find it difficult to garner valuable insights and make correct inferences. If any of these components are missing, you will not be able to realize an adequate return on your investment in the architecture.

A big data management architecture should be able to consume myriad data sources in a fast and inexpensive manner. Figure 2-1 outlines the architecture components that should be part of your *big data tech stack*. You can choose either open source frameworks or packaged licensed products to take full advantage of the functionality of the various components in the stack.

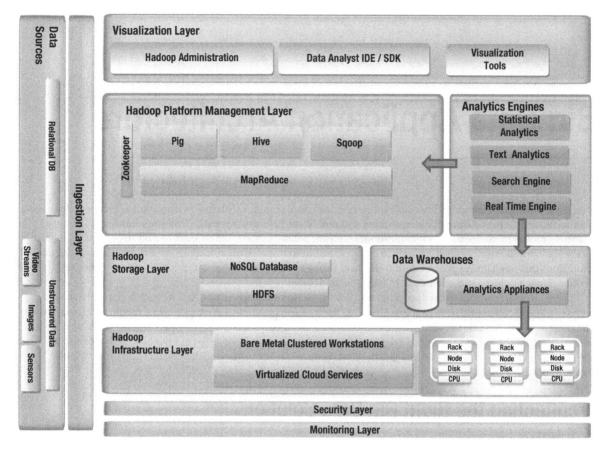

Figure 2-1. *The big data architecture*

Data Sources

Multiple internal and external data feeds are available to enterprises from various sources. It is very important that before you feed this data into your big data tech stack, you separate the noise from the relevant information. The signal-to-noise ratio is generally 10:90. This wide variety of data, coming in at a high velocity and in huge volumes, has to be seamlessly merged and consolidated later in the *big data stack* so that the *analytics* engines as well as the visualization tools can operate on it as one single big data set.

Problem

What are the various types of data sources inside and outside the enterprise that need to be analyzed in a big data solution? Can you illustrate with an industry example?

Solution

The real problem with defining *big data* begins in the *data sources layer*, where data sources of different volumes, velocity, and variety vie with each other to be included in the final big data set to be analyzed. These big data sets, also called *data lakes,* are pools of data that are *tagged* for inquiry or searched for patterns after they are stored in the Hadoop framework. Figure 2-2 illustrates the various types of data sources.

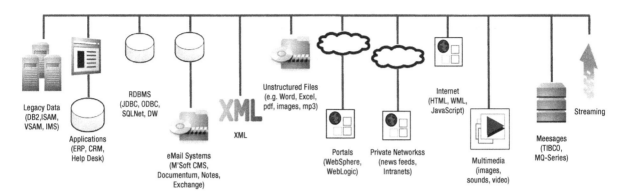

Figure 2-2. *The variety of data sources*

Industry Data

Traditionally, different industries designed their data-management architecture around the legacy data sources listed in Figure 2-3. The technologies, adapters, databases, and analytics tools were selected to serve these legacy protocols and standards.

Legacy Data Sources
HTTP/HTTPS web services
RDBMS
FTP
JMS/MQ based services
Text/flat file/csv logs
XML data sources
IM Protocol requests

Figure 2-3. *Legacy data sources*

In the past decade, every industry has seen an explosion in the amount of incoming data due to increases in subscriptions, audio data, mobile data, contentual details, social networking, meter data, weather data, mining data, devices data, and data usages. Some of the "new age" data sources that have seen an increase in volume, velocity, or variety are illustrated in Figure 2-4.

11

New Age Data Sources
High Volume Sources
1. Switching devices data
2. Access point data messages
3. Call data record due to exponential growth in user base
4. Feeds from social networking sites
Variety of Sources
1. Image and video feeds from social Networking sites
2. Transaction data
3. GPS data
4. Call center voice feeds
5. E-mail
6. SMS
High Velocity Sources
1. Call data records
2. Social networking site conversations
3. GPS data
4. Call center - voice-to-text feeds

Figure 2-4. *New age data sources—telecom industry*

All the data sources shown in Figure 2-4 have to be *funneled* into the enterprise after proper validation and cleansing. It is the job of the *ingestion layer* (described in the next section) to provide the functionality to be rapidly scalable for the huge inflow of data.

Ingestion Layer

The ingestion layer (Figure 2-5) is the new *data sentinel* of the enterprise. It is the responsibility of this layer to separate the noise from the relevant information. The ingestion layer should be able to handle the huge volume, high velocity, or variety of the data. It should have the capability to *validate, cleanse, transform, reduce,* and *integrate* the data into the *big data tech stack* for further processing. This is the new *edgeware* that needs to be *scalable, resilient, responsive,* and *regulatory* in the big data architecture. If the detailed architecture of this layer is not properly planned, the entire tech stack will be brittle and unstable as you introduce more and more capabilities onto your *big data analytics* framework.

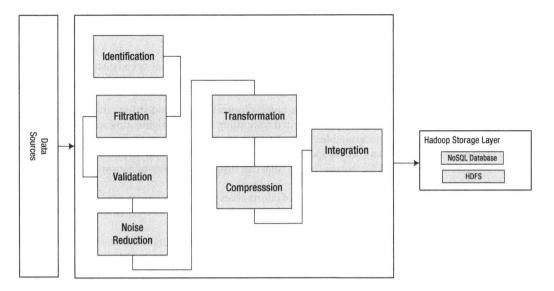

Figure 2-5. *Data ingestion layer*

Problem

What are the essential architecture components of the ingestion layer?

Solution

The ingestion layer loads the final relevant information, sans the noise, to the *distributed Hadoop storage layer* based on multiple commodity servers. It should have the capability to validate, cleanse, transform, reduce, and integrate the data into the big data tech stack for further processing.

The building blocks of the ingestion layer should include components for the following:

- *Identification* of the various known data formats or assignment of default formats to unstructured data.

- *Filtration* of inbound information relevant to the enterprise, based on the Enterprise MDM repository.

- *Validation* and analysis of data continuously against new MDM metadata.

- *Noise Reduction* involves cleansing data by removing the noise and minimizing distrurbances.

- *Transformation* can involve splitting, converging, denormalizing or summarizing data.

- *Compression* involves reducing the size of the data but not losing the relevance of the data in the process. It should not affect the analysis results after compression.

- *Integration* involves integrating the final massaged data set into the *Hadoop storage* layer— that is, Hadoop distributed file system (HDFS) and NoSQL databases.

There are multiple *ingestion patterns* (*data source-to-ingestion layer communication*) that can be implemented based on the performance, scalability, and availability requirements. Ingestion patterns are described in more detail in Chapter 3.

Distributed (Hadoop) Storage Layer

Using massively distributed storage and processing is a fundamental change in the way an enterprise handles big data. A distributed storage system promises fault-tolerance, and parallelization enables high-speed distributed processing algorithms to execute over large-scale data. The *Hadoop distributed file system* (HDFS) is the cornerstone of the big data storage layer.

Hadoop is an open source framework that allows us to store huge volumes of data in a distributed fashion across low cost machines. It provides de-coupling between the distributed computing software engineering and the actual application logic that you want to execute. Hadoop enables you to interact with a logical cluster of processing and storage nodes instead of interacting with the bare-metal operating system (OS) and CPU. Two major components of Hadoop exist: a massively scalable distributed file system (HDFS) that can support petabytes of data and a massively scalable *map reduce* engine that computes results in batch.

HDFS is a file system designed to store a very large volume of information (terabytes or petabytes) across a large number of machines in a cluster. It stores data reliably, runs on commodity hardware, uses blocks to store a file or parts of a file, and supports a write-once-read-many model of data access.

HDFS requires complex file read/write programs to be written by skilled developers. It is not accessible as a logical data structure for easy data manipulation. To facilitate that, you need to use new distributed, nonrelational data stores that are prevalent in the big data world, including key-value pair, document, graph, columnar, and geospatial databases. Collectively, these are referred to as *NoSQL*, or *not only SQL*, databases (Figure 2-6).

Figure 2-6. *NoSQL databases*

Problem

What are the different types of NoSQL databases, and what business problems are they suitable for?

Solution

Different NoSQL solutions are well suited for different business applications. Distributed NoSQL data-store solutions must relax guarantees around *consistency*, *availability,* and *partition tolerance* (the CAP Theorem), resulting in systems optimized for different combinations of these properties. The combination of relational and NoSQL databases ensures *the right data* is available when you need it. You also need data architectures that support complex unstructured content. Both relational databases and nonrelational databases have to be included in the approach to solve your big data problems.

Different NoSQL databases are well suited for different business applications as shown in Figure 2-7.

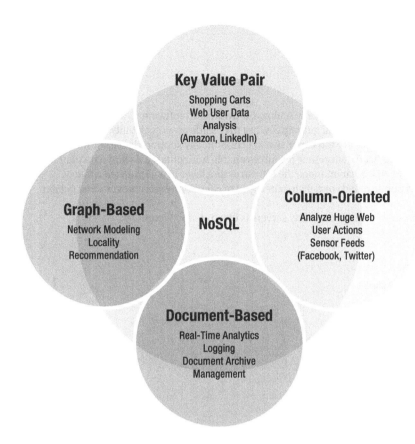

Figure 2-7. *NoSQL database typical business scenarios*

The storage layer is usually loaded with data using a batch process. The *integration* component of the *ingestion layer* invokes various mechanisms—like *Sqoop*, *MapReduce jobs*, *ETL jobs*, and others—to upload data to the *distributed Hadoop storage layer* (DHSL). The storage layer provides *storage patterns* (communication from ingestion layer to storage layer) that can be implemented based on the performance, scalability, and availability requirements. Storage patterns are described in more detail in Chapter 4.

Hadoop Infrastructure Layer

The layer supporting the strorage layer—that is, the physical infrastructure—is fundamental to the operation and scalability of big data architecture. In fact, the availability of a robust and inexpensive physical infrastructure has triggered the emergence of big data as such an important trend. To support unanticipated or unpredictable volume, velocity, or variety of data, a physical infrastructure for big data has to be different than that for traditional data.

The *Hadoop physical infrastructure layer* (HPIL) is based on a distributed computing model. This means that data can be physically stored in many different locations and linked together through networks and a distributed file system. It is a "share-nothing" architecture, where the data and the functions required to manipulate it reside together on a single node. Like in the traditional client server model, the data no longer needs to be transferred to a *monolithic server* where the SQL functions are applied to crunch it. Redundancy is built into this infrastructure because you are dealing with so much data from so many different sources.

Problem

What are the main components of a Hadoop infrastructure?

Solution

Traditional enterprise applications are built based on vertically scaling hardware and software. Traditional enterprise architectures are designed to provide strong transactional guarantees, but they trade away scalability and are expensive. Vertical-scaling enterprise architectures are too expensive to economically support dense computations over large scale data. Auto-provisioned, virtualized data center resources enable horizontal scaling of data platforms at significantly reduced prices. Hadoop and HDFS can manage the infrastructure layer in a *virtualized cloud environment* (on-premises as well as in a public cloud) or a distributed grid of *commodity servers* over a fast gigabit network.

A simple big data hardware configuration using commodity servers is illustrated in Figure 2-8.

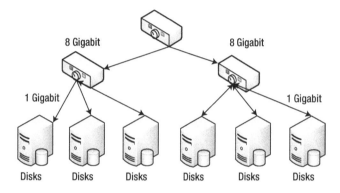

Figure 2-8. *Typical big data hardware topology*

The configuration pictured includes the following components: N commodity servers (8-core, 24 GBs RAM, 4 to 12 TBs, gig-E); 2-level network, 20 to 40 nodes per rack.

Hadoop Platform Management Layer

This is the layer that provides the tools and query languages to access the NoSQL databases using the HDFS storage file system sitting on top of the Hadoop physical infrastructure layer.

With the evolution of computing technology, it is now possible to manage immense volumes of data that previously could have been handled only by supercomputers at great expense. Prices of systems (CPU, RAM, and DISK) have dropped. As a result, new techniques for distributed computing have become mainstream.

Problem

What is the recommended data-access pattern for the Hadoop platform components to access the data in the Hadoop physical infrastructure layer?

Solution

Figure 2-9 shows how the platform layer of the big data tech stack communicates with the layers below it.

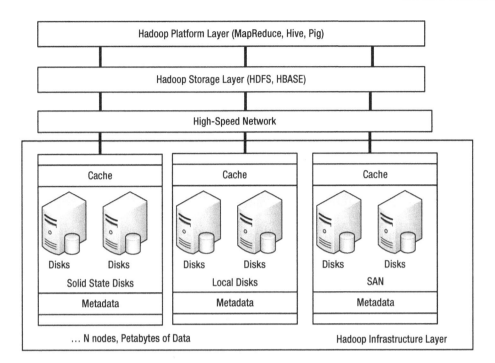

Figure 2-9. *Big data platform architecture*

Hadoop and MapReduce are the new technologies that allow enterprises to store, access, and analyze huge amounts of data in near real-time so that they can monetize the benefits of owning huge amounts of data. These technologies address one of the most fundamental problems—the capability to process massive amounts of data efficiently, cost-effectively, and in a timely fashion.

The *Hadoop platform management layer* accesses data, runs queries, and manages the lower layers using scripting languages like Pig and Hive. Various *data-access patterns* (communication from the platform layer to the storage layer) suitable for different application scenarios are implemented based on the performance, scalability, and availability requirements. Data-access patterns are described in more detail in Chapter 5.

Problem

What are the key building blocks of the Hadoop platform management layer?

Solution
MapReduce

MapReduce was adopted by Google for efficiently executing a set of functions against a large amount of data in batch mode. The *map* component distributes the problem or tasks across a large number of systems and handles the placement of the tasks in a way that distributes the load and manages recovery from failures. After the distributed computation is completed, another function called *reduce* combines all the elements back together to provide a result. An example of MapReduce usage is to determine the number of times big data has been used on all pages of this book. MapReduce simplifies the creation of processes that analyze large amounts of unstructured and structured data in parallel. Underlying hardware failures are handled transparently for user applications, providing a reliable and fault-tolerant capability.

Here are the key facts associated with the scenario in Figure 2-10.

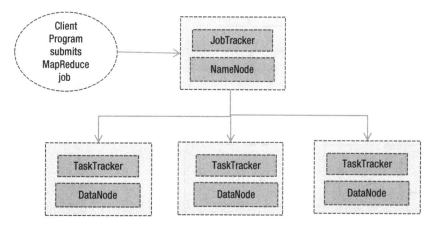

Figure 2-10. *MapReduce tasks*

- Each Hadoop node is part of an distributed cluster of machines cluster.

- Input data is stored in the HDFS distributed file system, spread across multiple machines and is copied to make the system redundant against failure of any one of the machines.

- The client program submits a batch job to the *job tracker*.

- The job tracker functions as the *master* that does the following:

  - Splits input data

  - Schedules and monitors various map and reduce tasks

- The task tracker processes are slaves that execute map and reduce tasks.

- *Hive* is a data-warehouse system for Hadoop that provides the capability to aggregate large volumes of data. This SQL-like interface increases the compression of stored data for improved storage-resource utilization without affecting access speed.

- *Pig* is a scripting language that allows us to manipulate the data in the HDFS in parallel. Its intuitive syntax simplifies the development of MapReduce jobs, providing an alternative programming language to Java. The development cycle for MapReduce jobs can be very long. To combat this, more sophisticated scripting languages have been created for exploring large datasets, such as Pig, and to process large datasets with minimal lines of code. Pig is designed for batch processing of data. It is not well suited to perform queries on only a small portion of the dataset because it is designed to scan the entire dataset.

- *HBase* is the column-oriented database that provides fast access to big data. The most common file system used with HBase is HDFS. It has no real indexes, supports automatic partitioning, scales linearly and automatically with new nodes. It is Hadoop compliant, fault tolerant, and suitable for batch processing.

- *Sqoop* is a command-line tool that enables importing individual tables, specific columns, or entire database files straight to the distributed file system or data warehouse (Figure 2-11). Results of analysis within MapReduce can then be exported to a relational database for consumption by other tools. Because many organizations continue to store valuable data in a relational database system, it will be crucial for these new NoSQL systems to integrate with relational database management systems (RDBMS) for effective analysis. Using extraction tools, such as Sqoop, relevant data can be pulled from the relational database and then processed using MapReduce or Hive, combining multiple datasets to get powerful results.

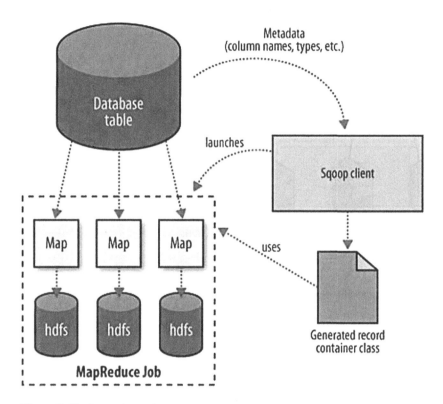

Figure 2-11. *Sqoop import process*

- *ZooKeeper* (Figure 2-12) is a coordinator for keeping the various Hadoop instances and nodes in sync and protected from the failure of any of the nodes. Coordination is crucial to handling partial failures in a distributed system. Coordinators, such as Zookeeper, use various tools to safely handle failure, including ordering, notifications, distributed queues, distributed locks, leader election among peers, as well as a repository of common coordination patterns. Reads are satisfied by followers, while writes are committed by the leader.

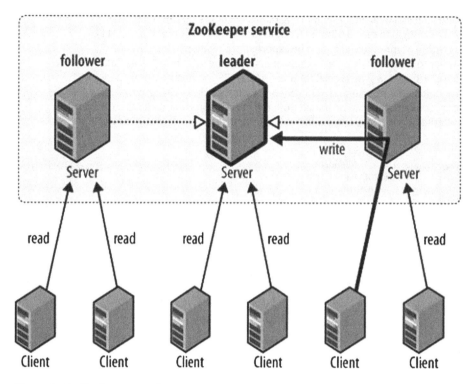

Figure 2-12. *Zookeeper topology*

Zookeeper guarantees the following qualities with regards to data consistency:

- Sequential consistency
- Atomicity
- Durability
- Single system image
- Timeliness

Security Layer

As big data analysis becomes a mainstream functionality for companies, security of that data becomes a prime concern. Customer shopping habits, patient medical histories, utility-bill trends, and demographic findings for genetic diseases—all these and many more types and uses of data need to be protected, both to meet compliance requirements and to protect the individual's privacy. Proper authorization and authentication methods have to be applied to the *analytics*. These security requirements have to be part of the big data fabric from the beginning and not an afterthought.

Problem

What are the basic security tenets that a big data architecture should follow?

Solution

An untrusted *mapper* or *named node job tracker* can return unwanted results that will generate incorrect reducer aggregate results. With large data sets, such security violations might go unnoticed and cause significant damage to the inferences and computations.

NoSQL injection is still in its infancy and an easy target for hackers. With large clusters utilized randomly for strings and archiving big data sets, it is very easy to lose track of where the data is stored or forget to erase data that's not required. Such data can fall into the wrong hands and pose a security threat to the enterprise.

Big data projects are inherently subject to security issues because of the distributed architecture, use of a simple programming model, and the open framework of services. However, security has to be implemented in a way that does not harm performance, scalability, or functionality, and it should be relatively simple to manage and maintain.

To implement a security baseline foundation, you should design a big data tech stack so that, at a minimum, it does the following:

- Authenticates nodes using protocols like *Kerberos*

- Enables file-layer encryption

- Subscribes to a key management service for trusted keys and certificates

- Uses tools like *Chef* or *Puppet* for validation during deployment of data sets or when applying patches on virtual nodes

- Logs the communication between nodes, and uses distributed logging mechanisms to trace any anomalies across layers

- Ensures all communication between nodes is secure—for example, by using Secure Sockets Layer (SSL), TLS, and so forth.

Monitoring Layer
Problem

With the distributed Hadoop grid architecture at its core, are there any tools that help to monitor all these moving parts?

Solution

With so many distributed data storage clusters and multiple data source ingestion points, it is important to get a complete picture of the big data tech stack so that the availability SLAs are met with minimum downtime.

Monitoring systems have to be aware of large distributed clusters that are deployed in a federated mode. The monitoring system has to be aware of different operating systems and hardware . . . hence the machines have to communicate to the monitoring tool via high level protocols like XML instead of binary formats that are machine dependent. The system should also provide tools for data storage and visualization. Performance is a key parameter to monitor so that there is very low overhead and high parallelism.

Open source tools like *Ganglia* and *Nagios* are widely used for monitoring big data tech stacks.

Analytics Engine
Co-Existence with Traditional BI

Enterprises need to adopt different approaches to solve different problems using big data; some analysis will use a traditional data warehouse, while other analysis will use both big data as well as traditional business intelligence methods.

The analytics can happen on both the data warehouse in the traditional way or on big data stores (using distributed MapReduce processing). Data warehouses will continue to manage RDBMS-based transactional data in a centralized environment. Hadoop-based tools will manage physically distributed unstructured data from various sources.

The mediation happens when data flows between the data warehouse and big data stores (for example, through Hive/Hbase) in either direction, as needed, using tools like Sqoop.

Real-time analysis can leverage low-latency NoSQL stores (for example, *Cassandra, Vertica*, and others) to analyze data produced by web-facing apps. Open source analytics software like *R* and *Madlib* have made this world of complex statistical algorithms easily accessible to developers and data scientists in all spheres of life.

Search Engines
Problem

Are the traditional search engines sufficient to search the huge volume and variety of data for finding the proverbial "needle in a haystack" in a big data environment?

Solution

For huge volumes of data to be analyzed, you need blazing-fast search engines with iterative and cognitive data-discovery mechanisms. The data loaded from various enterprise applications into the big data tech stack has to be indexed and searched for big data analytics processing. Typical searches won't be done only on database (HBase) rows (key), so using additional fields needs to be considered. Different types of data are generated in various industries, as seen in Figure 2-13.

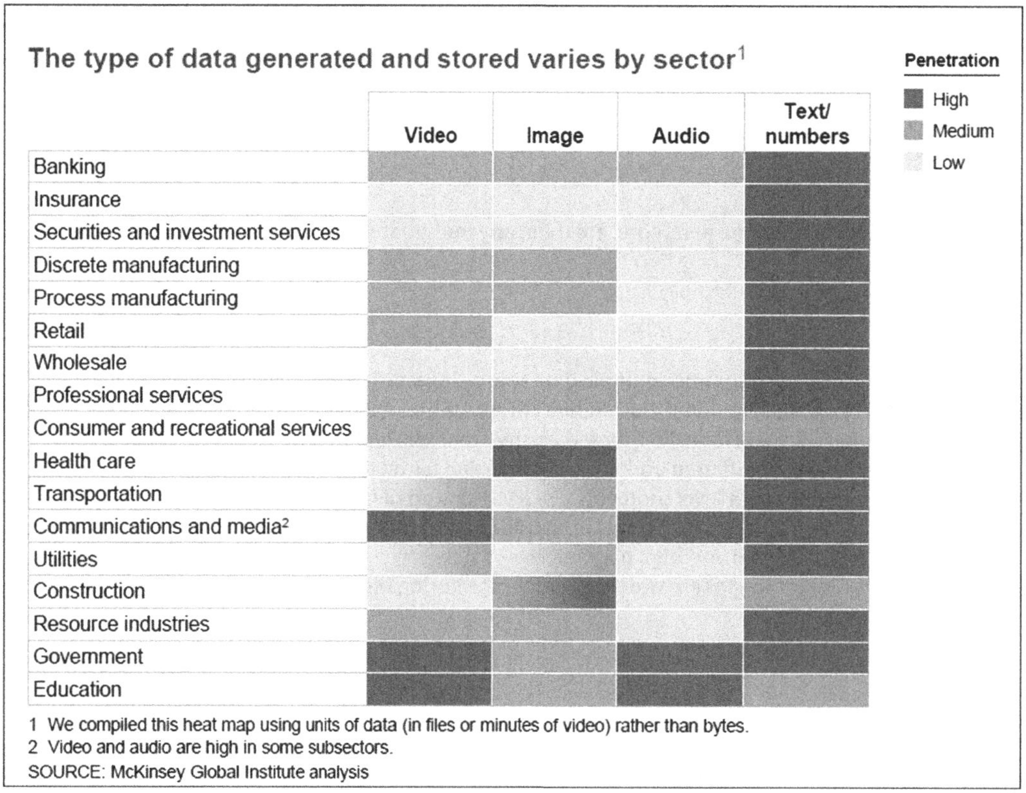

The type of data generated and stored varies by sector[1]

	Video	Image	Audio	Text/ numbers	Penetration
Banking					■ High
Insurance					▨ Medium
Securities and investment services					░ Low
Discrete manufacturing					
Process manufacturing					
Retail					
Wholesale					
Professional services					
Consumer and recreational services					
Health care					
Transportation					
Communications and media[2]					
Utilities					
Construction					
Resource industries					
Government					
Education					

1 We compiled this heat map using units of data (in files or minutes of video) rather than bytes.
2 Video and audio are high in some subsectors.
SOURCE: McKinsey Global Institute analysis

Figure 2-13. Search data types in various industries

Use of open source search engines like *Lucene*-based *Solr* give improved search capabilities that could serve as a set of secondary indices. While you're designing the architecture, you need to give serious consideration to this topic, which might require you to pick vendor-implemented search products (for example, *DataStax*). Search engine results can be presented in various forms using "new age" visualization tools and methods.

Figure 2-14 shows the conceptual architecture of the *search* layer and how it interacts with the various layers of a big data tech stack. We will look at *distributed search patterns* that meet the performance, scalability, and availability requirements of a big data stack in more detail in Chapter 3.

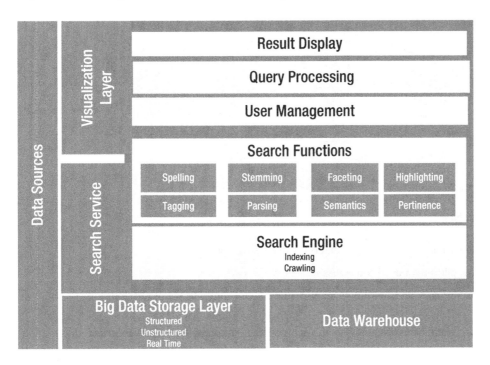

Figure 2-14. *Search engine conceptual architecture*

Real-Time Engines

Memory has become so inexpensive that pervasive visibility and real-time applications are more commonly used in cases where data changes frequently. It does not always make sense to store state to disk, using memory only to improve performance. The data is so humongous that it makes no sense to analyze it after a few weeks, as the data might be stale or the business advantage might have already been lost.

Problem

How do I analyze my data in real time for agile business intelligence capabilities in a big data environment?

Solution

To take advantage of the insights as early as possible, real-time options (where the most up-to-date data is found in memory, while the data on disk eventually catches up) are achievable using real-time engines and NoSQL data stores. Real-time analysis of web traffic also generates a large amount of data that is available only for a short period of time. This often produces data sets in which the schema is unknown in advance.

Document-based systems can send messages based on the incoming traffic and quickly move on to the next function. It is not necessary to wait for a response, as most of the messages are simple counter increments. The scale and speed of a NoSQL store will allow calculations to be made as the data is available. Two primary in-memory modes are possible for real-time processing:

- In-Memory Caching

  - Data is deployed between the application and the database to alleviate database load (Figure 2-15).

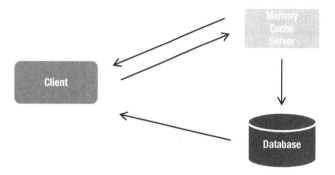

Figure 2-15. *In-memory caching*

- It's ideal for caching data to memory that is repeatedly accessed.

- Data is not replicated or persisted across servers.

- It harnesses the aggregate memory of many distributed machines by using a hashing algorithm.

To give you an example, Facebook uses 800 servers to supply over 28 TBs of memory for *Memcached*—for example, *Terracota*, *EHCache*.

- In-Memory Database

  - Data is deployed in the application tier as an embeddable database—for example, Derby (Figure 2-16).

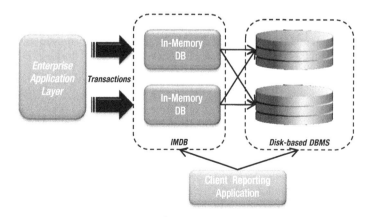

Figure 2-16. *In-memory database*

- Reading and writing data is as fast as accessing RAM. For example, with a 1.8-GHz processor, a read transaction can take less than 5 microseconds, with an insert transaction taking less than 15 microseconds.

- The database fits entirely in physical memory.

- The data is managed in memory with optimized access algorithms.

- Transaction logs and database checkpoint files are stored to disk.

Visualization Layer

Problem

Are the traditional analytical tools capable of interpreting and visualizing big data?

Solution

A huge volume of big data can lead to information overload. However, if visualization is incorporated early-on as an integral part of the big data tech stack, it will be useful for *data analysts and scientists* to gain insights faster and increase their ability to look at different aspects of the data in various visual modes.

Once the *big data* Hadoop processing aggregated output is *scooped* into the traditional ODS, data warehouse, and data marts for further analysis along with the transaction data, the visualization layers can work on top of this consolidated aggregated data. Additionally, if real-time insight is required, the real-time engines powered by *complex event processing* (CEP) engines and *event-driven architectures* (EDAs) can be utilized. Refer to Figure 2-17 for the interactions between different layers of the big data stack that allow you to harnesses the power of visualization tools.

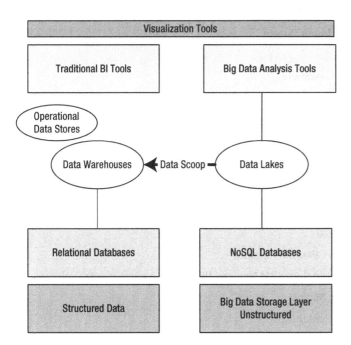

Figure 2-17. *Visualization conceptual architecture*

The *business intelligence layer* is now equipped with advanced big data analytics tools, in-database statistical analysis, and advanced visualization tools like *Tableau, Clickview, Spotfire, MapR, revolution R*, and others. These tools work on top of the traditional components such as reports, dashboards, and queries.

With this architecture, the business users see the traditional transaction data and big data in a consolidated single view. We will look at visualization patterns that provide agile and flexible insights into big data stacks in more detail in Chapter 6.

Big Data Applications

Problem

What is the minimum set of software tools I will need to implement, from end to end, the big data tech stack explained earlier?

Solution

You have a wide choice of tools and products you can use to build your application architecture end to end. We will look at many of them in later chapters as we discuss applying multiple-pattern scenarios to specific business scenarios. Products usually selected by many enterprises to begin their big data journey are shown in Table 2-1. The products listed are predominantly open source based, except for situations where an organization already has an IT investment in products from IBM, Oracle, SAP, EMC, and other companies and would like to leverage the existing licensing agreements to build big data environments at a reasonable price, as well as get continued support from the vendors.

Table 2-1. *Big data typical software stack*

Purpose	Products/tools
Ingestion Layer	Apache Flume, Storm
Hadoop Storage	HDFS
NoSQL Databases	Hbase, Cassandra
Rules Engines	MapReduce jobs
NoSQL Data Warehouse	Hive
Platform Management Query Tools	MapReduce, Pig, Hive
Search Engine	Solr
Platform Management Co-ordination Tools	ZooKeeper, Oozie
Analytics Engines	R, Pentaho
Visualization Tools	Tableau, Clickview, Spotfire
Big Data Analytics Appliances	EMC Greenplum, IBM Netezza, IBM Pure Systems, Oracle Exalytics
Monitoring	Ganglia, Nagios
Data Analyst IDE	Talend, Pentaho
Hadoop Administration	Cloudera, DataStax, Hortonworks, IBM Big Insights
Public Cloud-Based Virtual Infrastructure	Amazon AWS & S3, Rackspace

Problem

How do I transfer and load huge data into public, cloud-based Hadoop virtual clusters?

Solution

As much as enterprises would like to use the public cloud environments for their big data analytics, that desire is limited by the constraints in moving terabytes of data in and out of the cloud. Here are the traditional means of moving large data:

- Physically ship hard disk drives to a cloud provider. The risk is that they might get delayed or damaged in transit.

- The other digital means is to use *TCP-based* transfer methods such as *FTP* or *HTTP*.

Both options are woefully slow and insecure for fulfilling big data needs. To become a viable option for big data management, processing, and distribution, cloud services need a high-speed, non-TCP transport mechanism that addresses the bottlenecks of networks, such as the degradation in transfer speeds that occurs over distance using traditional transfer protocols and the last-mile loss of speed inside the cloud datacenter caused by the HTTP interfaces to the underlying object-based cloud storage.

There are products that offer better file-transfer speeds and larger file-size capabilities, like those offered by Aspera, Signiant, File catalyst, Telestream, and others. These products use a combination of *UDP protocol* and *parallel TCP validation*. UDP transfers are less dependable, and they verify by hash or just the file size, after the transfer is done.

Problem

Is Hadoop available only on Unix/Linux-based operating systems? What about Windows?

Solution

Hadoop is about commodity servers. More than 70 percent of the commodity servers in the world are Windows based. *Hortonworks data platform* (HDP) for Windows, a fully supported, open source Hadoop distribution that runs on Windows Server, was released in May 2013.

HDP for Windows is not the only way that Hadoop is coming to Windows. Microsoft has released its own distribution of Hadoop, which it calls *HDInsight*. This is available as a service running in an organization's Windows Azure cloud, or as a product that's intended to be used as the basis of an on-premises, private-cloud Hadoop installation.

Data analysts will be able to use tools like Microsoft Excel on HDP or HDInsight without the working through the learning curve that comes with implementing new visualization tools like Tableau and Clickview.

Summary

To venture into the *big data analytics* world, you need a robust architecture that takes care of visualization and real-time and offline analytics and is supported by a strong Hadoop-based platform. This is essential for the success of your program. You have multiple options when looking for products, frameworks, and tools that can be used to implement these logical components of the big data reference architecture. Having a holistic knowledge of these major components ensures there are no gaps in the planning phase of the architecture that get identified when you are halfway through your big data journey.

This chapter serves as the foundation for the rest of the book. Next we'll delve into the various interaction patterns across the different layers of the *big data architecture*.

▚ ▞ ▚

CHAPTER 3

Big Data Ingestion and Streaming Patterns

Traditional business intelligence (BI) and data warehouse (DW) solutions use structured data extensively. Database platforms such as Oracle, Informatica, and others had limited capabilities to handle and manage unstructured data such as text, media, video, and so forth, although they had a data type called CLOB and BLOB; which were used to store large amounts of text, and accessing data from these platforms was a problem. With the advent of multistructured (a.k.a. *unstructured*) data in the form of social media and audio/video, there has to be a change in the way data is ingested, preprocessed, validated, and/or cleansed and integrated or co-related with nontextual formats. This chapter deals with the following topics:

- How multistructured data is temporarily stored

- How the data integrity of large volumes can be maintained

- The physical taxonomy required to ensure fault-tolerant, streaming data ingestion

- Use of high-performance deployment patterns to ensure large volumes of data are ingested without any data loss

Understanding Data Ingestion

In typical ingestion scenarios, you have multiple data sources to process. As the number of data sources increases, the processing starts to become complicated. Also, in the case of big data, many times the source data structure itself is not known; hence, following the traditional data integration approaches creates difficulty in integrating data.

Common challenges encountered while ingesting several data sources include the following:

- Prioritizing each data source load

- Tagging and indexing ingested data

- Validating and cleansing the ingested data

- Transforming and compressing before ingestion

Problem

What are the typical data ingestion patterns?

Solution

Unstructured data, if stored in a relational database management system (RDBMS) will create performance and scalability concerns. Hence, in the big data world, data is loaded using multiple solutions and multiple target destinations to solve the specific types of problems encountered during ingestion.

Ingestion patterns describe solutions to commonly encountered problems in data source to ingestion layer communications. These solutions can be chosen based on the performance, scalability, and availability requirements. We'll look at these patterns (which are shown in Figure 3-1) in the subsequent sections. We will cover the following common data-ingestion and streaming patterns in this chapter:

- **Multisource Extractor Pattern:** This pattern is an approach to ingest multiple data source types in an efficient manner.

- **Protocol Converter Pattern:**–This pattern employs a protocol mediator to provide abstraction for the incoming data from the different protocol layers.

- **Multidestination Pattern:** This pattern is used in a scenario where the ingestion layer has to transport the data to multiple storage components like Hadoop Distributed File System (HDFS), data marts, or real-time analytics engines.

- **Just-in-Time Transformation Pattern:** Large quantities of unstructured data can be uploaded in a batch mode using traditional ETL (extract, transfer and load) tools and methods. However, the data is transformed only when required to save compute time.

- **Real-Time Streaming patterns:** Certain business problems require an instant analysis of data coming into the enterprise. In these circumstances, real-time ingestion and analysis of the in-streaming data is required.

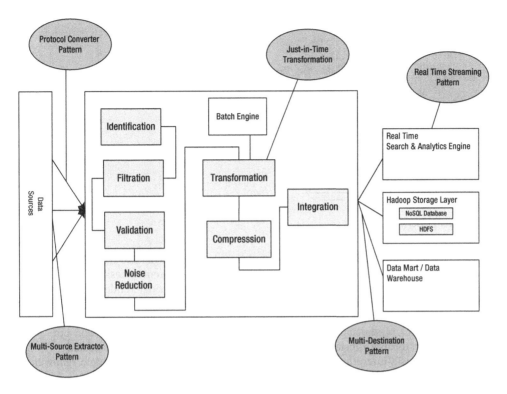

Figure 3-1. *Data ingestion layer and associated patterns*

Multisource Extractor Pattern
Problem

How will you ingest data from multiple sources and different formats in an efficient manner?

Solution

The multisource extractor pattern (shown in Figure 3-2) is applicable in scenarios where enterprises that have large collections of unstructured data need to investigate these disparate datasets and nonrelational databases (for example, NoSQL, Cassandra, and so forth); typical industry examples are claims and underwriting, financial trading, telecommunications, e-commerce, fraud detection, social media, gaming, and wagering. Feeds from energy exploration and video-surveillance equipment where application workloads are CPU and I/O-intensive are also ideal candidates for the multisource extractor pattern.

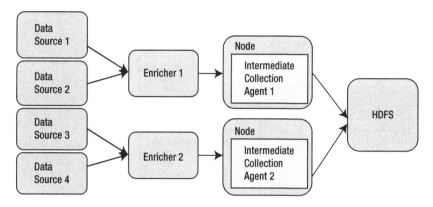

Figure 3-2. *Multisource extractor pattern*

Multisource extractor taxonomy ensures that the ingestion tool/framework is highly available and distributed. It also ensures that huge volumes of data get segregated into multiple batches across different nodes. For a very small implementation involving a handful of clients and/or only a small volume of data, even a single-node implementation will work. But, for a continuous stream of data influx from multiple clients and a huge volume, it makes sense to have clustered implementation with batches partitioned into small volumes.

Generally, in large ingestion systems, big data operators employ *enrichers* to do initial data aggregation and cleansing. (See Figure 3-2.) An *enricher* reliably transfers files, validates them, reduces noise, compresses and transforms from a native format to an easily interpreted representation. Initial data cleansing (for example, removing duplication) is also commonly performed in the enricher tier.

Once the files are processed by enrichers, they are transferred to a cluster of *intermediate collectors* for final processing and loading to destination systems.

Because the ingestion layer has to be fault-tolerant, it always makes sense to have multiple nodes. The number of disks and disk size per node have to be based on each client's volume. Multiple nodes will be able to write to more drives in parallel and provide greater throughput.

However, the multisource extractor pattern has a number of significant disadvantages that make it unusable for real-time ingestion. The major shortcomings are as follows:

- **Not Real Time:** Data-ingestion latency might vary between 30 minutes and a few hours.

- **Redundant Data:** Multiple copies of data need to be kept in different tiers of enrichers and collection agents. This makes already large data volumes even larger.

- **High Costs:** High availability is usually a requirement for this pattern. As the systems grow in capacity, costs of maintaining high availability increases.

- **Complex Configuration:** This batch-oriented pattern is difficult to configure and maintain.

Table 3-1 outlines a sample textual data ingestion using a single-node taxonomy against a multinode taxonomy.

Table 3-1. *Distributed and Clustered Flume Taxonomy*

	Time to Ingest 2 TB	Disk size/ Node	No. of disks / Node	RAM
Single Node Collector	3.5 Hours	1 GB/disk	4	4GB
2- Node Collector	1 Hour	1 GB/disk	4	4 GB
Single-Node Collector	3.5 hours	1 GB/disk	4	4 GBs
2-Node Collector	1 hour	1 GB/Disk	4	4 GBs

Protocol Converter Pattern

Problem

How will you ingest data from multiple sources and different formats/protocols in an efficient manner?

Solution

The protocol converter pattern (shown in Figure 3-3) is applicable in scenarios where enterprises have a wide variety of unstructured data from data sources that have different data protocols and formats. In this pattern, the ingestion layer does the following:

1. Identifies multiple channels of incoming event.

2. Identifies polydata structures.

3. Provides services to mediate multiple protocols into suitable sinks.

4. Provides services to interface binding of external system containing several sets of messaging patterns into a common platform.

5. Provides services to handle various request types.

6. Provides services to abstract incoming data from various protocol layers.

7. Provides a unifying platform for the next layer to process the incoming data.

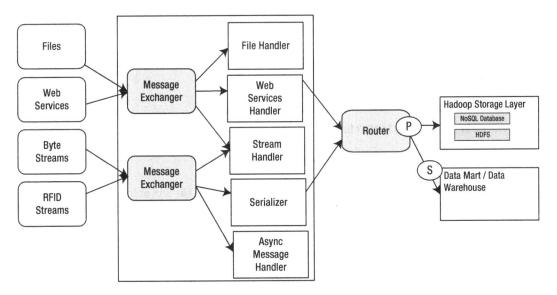

Figure 3-3. *Protocol converter pattern*

Protocol conversion is required when the source of data follows various different protocols. The variation in the protocol is either in the headers or the actual message. It could be either the number of bits in the headers, the length of the various fields and the corresponding logic required to decipher the data content, the message could be fixed length or variable length with separators.

This pattern is required to standardize the structure of the various different messages so that it is possible to analyze the information together using an analytics tool . The converter fits the different messages into a standard canonical message format that is usually mapped to a NoSQL data structure.

This concept is important when a system needs to be designed to address multiple protocols having multiple structures for incoming data.

In this pattern, the ingestion layer provides the following services:

- **Message Exchanger:** The messages could be synchronous or asynchronous depending on the protocol used for transport. A typical example is a web application information exchange over HTPP and the JMS-like message oriented communication that is usually asynchronous.

- **Stream Handler:** This component recognizes and transforms data being sent as byte streams or object streams—for example, bytes of image data, PDFs, and so forth.

- **File handler:** This component recognizes and loads data being sent as files—for example, FTP.

- **Web Services Handler:** This component defines the manner of data population and parsing and translation of the incoming data into the agreed-upon format—for example, REST WS, SOAP-based WS, and so forth.

- **Async Handler:** This component defines the system used to handle asynchronous events—for example, MQ, Async HTTP, and so forth.

- **Serializer:** The serializer handles incoming data as *Objects* or complex types over RMI (remote method invocation)—for example, EJB components. The object state is stored in databases or *flat files*.

Multidestination Pattern
Problem

Should all the raw data be ingested only in HDFS? In what scenario should it be ingested in multiple destinations?

Solution

Many organizations have traditional RDBMS systems as well as analytics platforms like SAS or Informatica. However, the ever-growing amount of data from an increasing number of data streams causes storage overflow problems. Also, the cost of licenses required to process this huge data slowly starts to become prohibitive. Increasing volume also causes data errors (a.k.a., *data regret*), and the time required to process the data increases exponentially. Because the RDBMS and analytics platforms are physically separate, a huge amount of data needs to be transferred over the network on a daily basis.

To overcome these challenges, an organization can start ingesting data into multiple data stores, both RDBMS as well as NoSQL data stores. The data transformation can be performed in the HDFS storage. Hive or Pig can be used to analyze the data at a lower cost. This also reduces the load on the existing *SAS/Informatica analytics* engines.

The Hadoop layer uses *map reduce* jobs to prepare the data for effective querying by *Hive* and *Pig*. This also ensures that large amounts of data need not be transferred over the network, thus avoiding huge costs.

The multidestination pattern (Figure 3-4) is very similar to the multisource ingestion pattern until it is ready to integrate with multiple destinations. A router publishes the "enriched" data and then broadcasts it to the subscriber destinations. The destinations have to register with the publishing agent on the router. Enrichers can be used as required by the publishers as well as the subscribers. The router can be deployed in a cluster, depending on the volume of data and number of subscribing destinations.

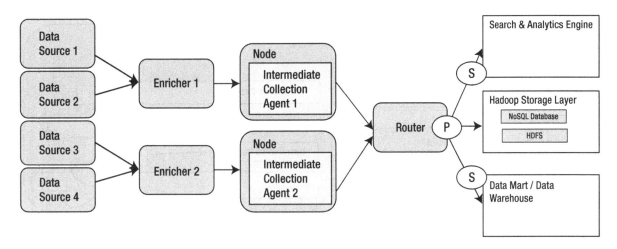

Figure 3-4. *Multidestination pattern*

This pattern solves some of the problems of ingesting and storing huge volumes of data:

- Splits the cost of storage by dividing stored data among traditional storage systems and HDFS.

- Provides the ability to partition the data for flexible access and processing in a decentralized fashion.

- Due to replication on the HDFS nodes, there is no "data regret."

- Because each node is self-sufficient, it's easy to add more nodes and storage without delays.

- Decentralized computation at the data nodes without extraction of data to other tools.

- Allows use of simple query languages like Hive and Pig alongside the giants of traditional analytics.

Just-in-Time Transformation Pattern
Problem

Should preprocessing of data—for example, cleansing/validation—always be done before ingesting data in HDFS?

Solution

For a huge volume of data and a huge number of analytical computations, it makes sense to ingest all the raw data into HDFS and then run dependent preprocessing batch jobs based on the business case to be implemented to cleanse, validate, co-relate, and transform the data. This transformed data, then, can again be stored in HDFS itself or transferred to data marts, warehouses, or real-time analytics engines. In short, raw data and transformed data can co-exist in HDFS and running all preprocessing transformations before ingestion might not be always ideal.

But basic validations can be performed as part of preprocessing on data being ingested.

This section introduces you to the *just-in-time* transformation pattern, where data is loaded and then transformed when required by the business. Notice the absence of the enricher layer in Figure 3-5. Multiple batch jobs run in parallel to transform data as required in the HDFS storage.

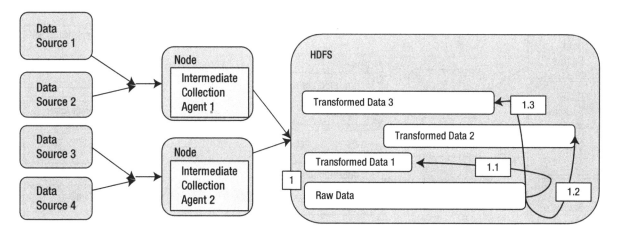

Figure 3-5. *Raw data as well as transformed data co-existing in HDFS*

Real-Time Streaming Pattern

Problem

How do we develop big data applications for processing continuous, real-time and unstructured inflow of data into the enterprise?

Solution

The key characteristics of a real-time streaming ingestion system (Figure 3-6) are as follows:

- It should be self-sufficient and use local memory in each processing node to minimize latency.

- It should have a share-nothing architecture—that is, all nodes should have atomic responsibilities and should not be dependent on each other. .

- It should provide a simple API for parsing the real time information quickly.

- The atomicity of each of the components should be such that the system can scale across clusters using commodity hardware.

- There should be no centralized master node. All nodes should be deployable with a uniform script.

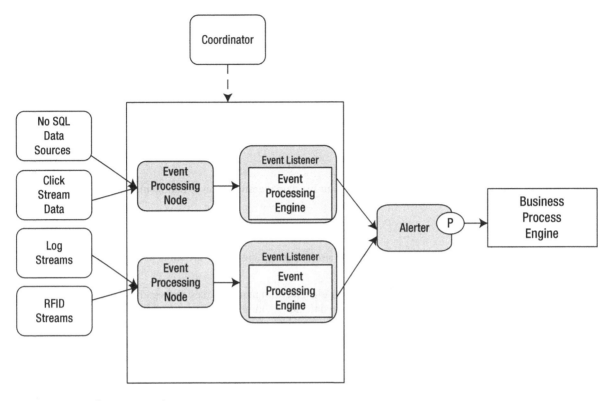

Figure 3-6. *Real-time streaming pattern*

Event processing nodes (EPs) consume the different inputs from various data sources. EPs create events that are captured by the event listeners of the *event processing engines*. Event listeners are the logical hosts to EPs. Event processing engines have a very large in-memory capacity (*big memory*). EPs get triggered by events as they are based on an event driven architecture. As soon as a event occurs the EP is triggered to execute a specific operation and then forward it to the alerter. The *alerter* publishes the results of the in-memory big data analytics to the enterprise BPM (business process management) engines. The BPM processes can redirect the results of the analysis to various channels like mobile, CIO dashboards, BAM systems and so forth.

Problem

What are the essential tools/frameworks required in your big data ingestion layer to handle files in batch-processing mode?

Solution

There are many product options to facilitate batch-processing-based ingestion. Here are some of the major frameworks available in the market:

- *Apache Sqoop* is a is used to transfer large volumes of data between Hadoop big data nodes and relational databases.. It offers two-way replication with both snapshots and incremental updates.

- *Chukwa* is a Hadoop subproject that is designed for efficient log processing. It provides a scalable distributed system for monitoring and analysis of log-based data. It supports appending to existing files and can be configured to monitor and process logs that are generated incrementally across many machines.

- *Apache Kafka* is a broadcast messaging system where the information is being listened to by multiple subscribers and picked up based on relevance to each subscriber. The publisher can be configured to retain the messages until the confirmation is received from all the subscribers. If any subscriber does not receive the information, the publisher will send it again. Its features include the use of compression to optimize IO performance and mirroring to improve availability, improve scalability, to optimize performance in multiple-cluster scenarios. It can be used as the framework between the router and Hadoop in the multidestination pattern implementation.

Problem

What are the essential tools/frameworks required in your big data ingestion layer to handle real-time streaming data?

Solution

There are many product options to facilitate real-time streaming ingestion. Here are some of the major frameworks available in the market:

- *Flume* is a distributed system for collecting log data from many sources, aggregating it, and writing it to HDFS. It is based on streaming data flows. Flume provides extensibility for online analytic applications. However, Flume requires a fair amount of configuration that can become very complex for very large systems.

- *Storm* supports event-stream processing and can respond to individual events within a reasonable time frame. Storm is a general-purpose, event-processing system that uses a cluster of services for scalability and reliability. In Storm terminology, you create a topology that runs continuously over a stream of incoming data. The data sources for the topology are called *spouts*, and each processing node is called a *bolt*. Bolts can perform sophisticated computations on the data, including output to data stores and other services. It is common for organizations to run a combination of Hadoop and Storm services to gain the best features of both platforms.

- *InfoSphere Streams* is able to perform complex analytics of heterogeneous data types. Infosphere Streams can support all data types. It can perform real-time and look-ahead analysis of regularly generated data, using digital filtering, pattern/correlation analysis, and decomposition as well as geospatial analysis. *Apache S4 is a Yahoo invented platform for handling continuous real time ingestion of data. It provides simple APIs for manipulating the unstructured streams of data, searches and distributes the processing across multiple nodes automatically without complicated programming.* Client programs that send and receive events can be written in any programming language. S4 is *designe*d as a highly distributed system. Throughput can be increased linearly by adding nodes into a cluster. The S4 design is best suited for large-scale applications for data mining and machine learning in a production environment.

ETL Tools for Big Data

Problem

Can traditional ETL tools be used to ingest data into HDFS?

Solution

Traditional ETL tools like the *open source Talend, Pentahho DI*, or well-known products like *Informatica* can also be leveraged for *data ingestion*. Some of the traditional ETL tools can read and write multiple files in parallel from and to HDFS.

ETL tools help to get data from one data environment and put it into another data environment. ETL is generally used with batch processing in data warehouse environments. Data warehouses provide business users with a way to consolidate information across disparate sources to analyze and report on insights relevant to their specific business focus. ETL tools are used to transform the data into the format required by the data warehouse. The transformation is actually done in an intermediate location before the data is loaded into the data warehouse.

In the big data world, ETL tools like Informatica have been used to enable a fast and flexible ingestion solution (greater than 150 GBs/day) that can support ad hoc capability for data and insight discovery. Informatica can be used in place of *Sqoop* and *Flume* solutions. *Informatica PowerCenter* can be utilized as a primary raw data ingestion engine.

Figure 3-7 depicts a scenario in which a traditional ETL tool has been used to ingest data into HDFS.

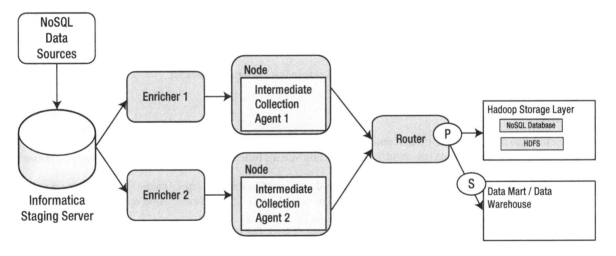

Figure 3-7. *Ingestion using traditional ETL tools*

Problem

Are there message-transformation best practices in the ingestion layers that facilitate faster ingestion?

Solution

For the transformation of messages and location of the transformation process, these guidelines can be followed:

- Perform the transformation as a part of the extraction process.

This allows selecting only the record of interest for loading and, ideally, that should be the data that has changed since the last extraction. Simple transformation, such as decoding an attribute and uppercase/lowercase conversions, can be performed in the source system.

- Perform the transformation as a separate layer.

Transformation can be performed in the staging area prior to loading in the HDFS system. When there are multiple data sources, the data needs to be consolidated and mapped to a different data structure. Such an intermediate data store is called *staging*. We can use Hadoop ETL tools like Hive and Pig in this area to do the transformation.

- Perform the transformation as a part of the loading process.

Some simple transformation can be done during the loading process itself.

- Perform the transformation in memory.

Transformation of data in memory is considered to be a better option for large and complex transformations with no latency between the extraction and loading processes. But this involves large amounts of memory. This is useful for near real-time system analytics (for example, SAP HANA), where transformation and loading is done with very low latency.

Problem

Are there any hardware appliances available in the market to ingest big data workloads?

Solution

Many vendors like IBM, Oracle, EMC, and others have come out with hardware appliances that promise end-to-end systems that optimize the big data ingestion, processing, and storage functions. The next section gives you a brief idea about the capabilities of such appliances.

Oracle has a big data appliance that handles data to the tune of 600 TBs. It utilizes Apache CDH for big data management. It also has an inherent 10 GbE high speed network between the nodes for rapid real time ingestion and replication.

EMC comes with a similar appliance called Greenplum with similar features to facilitate high speed low cost data processing using Hadoop.

Problem

How do I ingest data onto third-party public cloud options like Google BigQuery?

Solution

BigQuery is Google's cloud-based big data analytics service. You can upload your big data set to BigQuery for analysis. However, depending on your data's structure, you might need to prepare the data before loading it into BigQuery. For example, you might need to export your data into a different format or transform the data. BigQuery supports two data formats: CSV and JSON.

BigQuery can load uncompressed files significantly faster than compressed files due to parallel load operations, but because uncompressed files are larger in size, using them can lead to bandwidth limitations and higher Google Cloud Storage costs. For example, uncompressed files that live on third-party services can consume considerable bandwidth and time if uploaded to Google Cloud Storage for loading.

With BigQuery, processing is faster if you denormalize the data structure to enable super-fast querying.

Large datasets are often represented using XML. BigQuery doesn't support directly loading XML files, but XML files can be easily converted to an equivalent JSON format or flat CSV structure and then uploaded to continue processing.

Summary

With the huge volume of data coming into enterprises from various data sources, different challenges are encountered that can be solved using the patterns mentioned in this chapter. These patterns provide topologies to address multiple types of data formats and protocols, as well provide guidance about how much time it takes to process the data, the location of the transformation and so forth. A judicious use of these patterns will help the big data architect sift the noise from the true information before it enters the enterprise for further analysis in the Hadoop storage area.

CHAPTER 4

Big Data Storage Patterns

There are various storage infrastructure options available in the market, and big data appliances have added a new dimension to infrastructure options. Enterprises can leverage their existing infrastructure and storage licenses in addition to these new solutions for big data. In this chapter, we will cover the various storage mechanisms available, as well as patterns that amalgamate existing application storage frameworks with new big data implementations.

Understanding Big Data Storage

Since data is now more than just plain text, it can exist in various persistence-storage mechanisms, with Hadoop distributed file system (HDFS) being one of them. The way data is ingested or the sources from which data is ingested affects the way data is stored. On the other hand, how the data is pushed further into the downstream systems or accessed by the data access layer decides how the data is to be stored.

The need to store huge volumes of data has forced databases to follow new rules of data relationships and integrity that are different from those of relational database management systems (RDBMS).

RDBMS follow the ACID rules of atomicity, consistency, isolation and durability. These rules make the database reliable to any user of the database. However, searching huge volumes of big data and retrieving data from them would take large amounts of time if all the ACID rules were enforced.

A typical scenario is when we search for a certain topic using Google. The search returns innumerable pages of data; however, only one page is visible or *basically available* (BA). The rest of the data is in a *soft* state (S) and is still being assembled by Google, though the user is not aware of it. By the time the user looks at the data on the first page, the rest of the data becomes *eventually* consistent (E). This phenomenon—basically available soft state and eventually consistent—is the rule followed by the big data databases, which are generally NoSQL databases following BASE properties.

Database theory suggests that any distributed NoSQL big database can satisfy only two properties predominantly and will have to relax standards on the third. The three properties are *consistency*, *availability*, and *partition* tolerance (CAP). This is the CAP theorem.

The aforementioned paradigms of ACID, BASE, and CAP give rise to new big data storage patterns (Figure 4-1) like the following:

- **Façade pattern:** HDFS serves as the intermittent façade for the traditional DW systems.

- **Lean pattern:** HBase is indexed using only one column-family and only one column and unique row-key.

- **NoSQL pattern:** Traditional RDBMS systems are replaced by NoSQL alternatives to facilitate faster access and querying of big data.

- **Polyglot pattern:** Multiple types of storage mechanisms—like *RDBMS, file storage, CMS, OODBMS, NoSQL* and *HDFS*—co-exist in an enterprise to solve the big data problem.

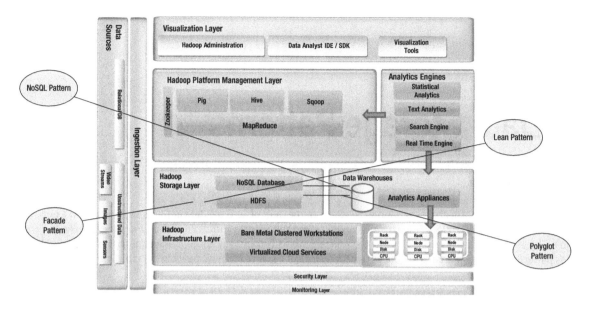

Figure 4-1. *Big data storage patterns*

Façade Pattern
Problem

Does big data replace existing data warehouse (DW) implementations?

Solution

Hadoop is not necessarily a replacement for a *data warehouse* (DW). It can also act as façade for a DW (Figure 4-2 and Figure 4-3). Data from different sources can be aggregated into an HDFS before being transformed and loaded to the traditional DW and business intelligence (BI) tools.

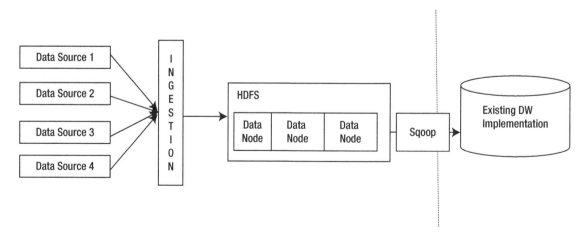

Figure 4-2. *The Hadoop Façade pattern*

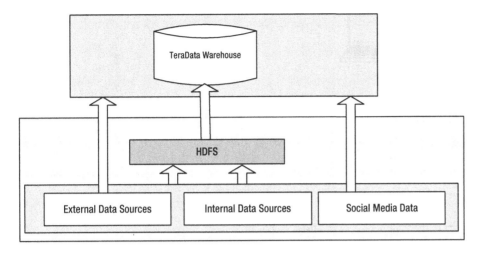

Figure 4-3. *Hadoop as a façade for TeraData*

This helps in retaining the investment in the existing DW framework, as well as the data usage in the downstream systems. This also helps re-use the existng infrastructure and add an abstraction of the DW. Hence, if new data sources are added to the ingestion system, it is still abstracted from the DW framework. This pattern solves the *variety* challenge among the three Vs (velocity, variety, and volume) of big data as shown in the example in Figures 4-3 to 4-6.

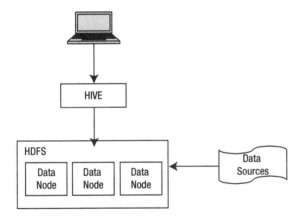

Figure 4-4. *Typical big data storage and access*

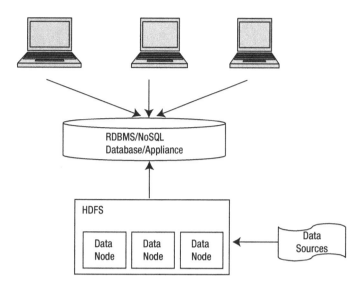

Figure 4-5. *Abstraction of RDBMS above HDFS*

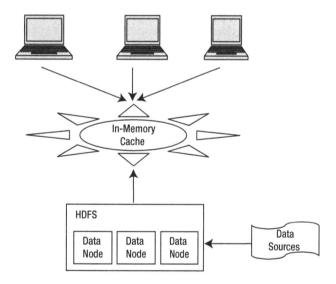

Figure 4-6. *Abstraction of in-memory cache above HDFS*

Data can be stored as "structured" data after being ingested into HDFS in the form of storage in an RDBMS (Figure 4-5) or in the form of applicances like *IBM Netezza/EMC Greenplum*, NoSQL Databases like *Cassandra/HO Vertica/Oracle Exadata*, or simply in an in-memory cache (Figure 4-6).

This ensures that only necessary data resides in the "structured" storage, thereby reducing the data size as well as latency (while accessing the data).

Data Appliances

The types of big data appliances claiming to offer high performance and low latency have mushroomed in the market. We need to be aware how they really affect the existing infrastructure and the benefits that these appliances bring to the table.

Problem

What are the benefits of using an integrated big data appliance?

Solution

HP Vertica, IBM Netezza, Oracle Exadata, and EMC Greenplum are packaged, commercial off-the-shelf COTS appliances available in the market. The advantage of such appliances is that they bring together the infrastructure, the Hadoop firmware, and management tools (for managing the Hadoop nodes). These appliances also ensure that instead of aligning with multiple vendors for software, storage and tools, only a tie-up with a single vendor is needed. This reduces considerable legwork for the client, ensuring the client deals with a single vendor for all issues.

HP Vertica (Figure 4-7) is an example of an all-in-one appliance.

Entity	Vendor tie-ups
Hadoop Software Distribution	Cloudera, Hortonworks or MAPR distribution
Storage	HP Vertica – RAID compliant columnar database
Infrastructure	HP Proliant servers
Analytics/Visualization	SAS
Machine Learning	R

Figure 4-7. *Vendors with multiple tie-ups*

As you can see from the preceding example, the HP big data implementation brings with it back-to-back collaboration with other vendors. Other examples are shown in Figure 4-8.

Vendor	What the vendor brings to the table
EMC	Greenplum appliance + EMC Storage HW + Pivotal HD (Hadoop Distribution) + TeraData
Oracle	Exadata + Sun SPARC servers + Exalytics
IBM	Big Insights + Netezza + PureData + PureSystems

Figure 4-8. *List of big data vendors with multiple tie-ups*

For one financial firm, a shift from a relational database to an appliance saw a 270 times faster query run. So the difference can be quite substantial.

The points to consider before implementing an appliance are these:

- Vendor lock-in

- Time to bring appliance implementation into production (porting data from legacy applications to appliance)

- Skills and expertise availability

- License cost

- Annual maintenance cost

- Total cost of ownership (TCO)

- Return on investment (ROI)

- Business case or the need for performance improvements over the existing Hadoop implementation

- Security (encryption and authentication)

- Integration with existing tools and hardware

Storage Disks

SAN, NAS, and SSD are some well-known storage formats. Big data has been tested on SAN disks, but there is not much performance data available regarding SSD.

Problem

Should big data be stored on RAID-configured disks?

Solution

RAID configuration is not necessary if the default storage is HDFS, because it already has a replication mechanism. Some appliances, including some of those discussed earlier, abstract the data that needs to be analyzed into a "structured" format that might have to be RAID-configured.

Data Archive/Purge
Problem

Is there a time-to-live for data to reside in HDFS?

Solution

Yes, as in any storage solution, data needs to be persisted only as long as the business demands it. Beyond this period, the data can be archived to products like *EMC Isilon* (`http://www.emc.com/archiving/archive-big-data.htm`). A data purge also has to be business driven.

Data Partitioning/Indexing and the Lean Pattern

Problem

HDFS is a distributed file system and inherently splits a file into chunks and replicates them. Does it still need further partitioning and indexing?

Solution

As a best practice, data partitioning is recommended for HDFS-based NoSQL databases. Because HDFS is a folder structure, data can be distributed/partitioned in multiple folders that are created on the basis of time-stamp, geography, or any other parameters that drive the business. This ensures that data access is capable of very high performance.

Problem

Is data indexing possible in HDFS (the Lean pattern)?

Solution

Data indexing as known in the RDBMS world is not applicable to HDFS but is applicable to NoSQL databases, like HBase, that are HDFS aware and compliant.

HBase works on the concept of column-family apart from columns, which can be leveraged to aggregate similar data together (Figures 4-9 and 4-10).

Column Family			
Column1	**Column2**	**Column3**	**Column4**

Figure 4-9. *HBase implementation with only one column-family and multiple columns*

Column Family 1			Column Family 2		
Column1	Column2	Column3	Column11	Column12	Column13

Figure 4-10. *HBase implementation with multiple column-families and multiple columns*

As you can see in the preceding illustrations, there are three ways a dataset can be uniquely identified. A unique combination of column-family name and a column can be used to uniquely identify a dataset. This can be achieved by having a combination of one-column or multiple-column families. A third way is to create a unique row-key, while having only a one column-family and one column. This implementation is called a ***Lean pattern*** implementation (Figure 4-11). The row-key name should end with a suffix of a time-stamp.

	Column Family
Row-Key	Column

Figure 4-11. *Lean pattern—HBase implementation with only one column-family and only one column and unique row-key*

This not only helps create a unique row-key but also helps in filtering or sorting data because the suffix is numeric in the form of a time-stamp.

Since maintenance can be difficult if the Lean pattern is implemented, it should be chosen over the other two only if the right skills and expertise exist in the big data team.

HDFS Alternatives
Problem

Are there other publicly available big data storage mechanisms?

Solution

Amazon Web Services (AWS) has its own storage mechanism in the form of S3. Data can be stored in S3 in the form of buckets. Whether all the data resides in a single bucket or in multiple buckets, again, should be driven by business needs and/or the skills available in the organization.

Other vendors, like IBM (GPFS) and EMC (Figure 4-12), have also been marketing their own file systems, but not many industry credentials are present to make them serious contenders to HDFS.

Vendor	Alternative
Amazon	S3
IBM	GPFS
EMC	Isilon OneFS
MapR	MapR file system

Figure 4-12. *HDFS alternatives*

MapR claims to have a file system two times faster than HDFS. (See http://www.slideshare.net/mcsrivas/design-scale-and-performance-of-maprs-distribution-for-hadoop.) However, clients would be reluctant to have a vendor lock-in at the file level. Migrating from a MapR Hadoop distribution to a Cloudera or a Hortonworks distribution will surely result in different performance statistics.

NoSQL Pattern
Problem

What role do NoSQL databases play in the Hadoop implementation?

Solution

NoSQL databases can store data on local NFS disks as well as HDFS. NoSQL databases are HDFS-aware; hence, data can be distributed across Hadoop data nodes and, at the same time, data can be easily accessed because it is stored in a nonrelational, columnar fashion. As we have discussed there are four types of NoSQL databases. Figure 4-13 lists their major big data use cases. Vendor implementations of NoSQL subsequently became open source implementations as seen in Figure 4-14.

NoSQL DB to Use	Scenario
Graph Database	Applications that provide evaluations of "like" or note that "user that bought this item also bought," like a recommendation engine.
Key-Value Pair Database	Needle-in-a-haystack applications.
Document Database	Applications that evaluate churn management on the basis of non-enterprise and social media data.
Columnar Database	Google search type of applications, where an entire related columnar family needs to be retrieved based on a string.

Figure 4-13. *NoSQL use cases*

Company	Product	OSS Equivalent	Most widely Used by
Amazon	DynamoDB	M/DB,	LinkedIn, Mozilla
Facebook	Cassandra	Cassandra	NetFlix, Twitter
Google	BigTable	HBase	Adobe Photoshop

Figure 4-14. *NoSQL vendors*

Scenarios for NoSQL:

- "N+1 Selects" problem for a large dataset

- Writing a huge amount of data

- Semi structured data

You should investigate NoSQL technologies to determine which offerings best fit your needs.

As mentioned, NoSQL databases allow faster searching on huge unstructured data.

Key-value pair databases store data as simple key-value pairs. The keys are unique and do not have any foreign keys or constraints. They are suitable for parallel lookups because the data sources have no relationships among each other. As you can imagine, such a structure is good for high read access. Due to a lack of referential integrity, the data integrity has to be managed by the front-end applications.

Column-oriented databases have a huge number of columns for each tuple. Each column also has a column key. Related columns have a column-family qualifier so that they can be retrieved together during a search. Because each column also has a column key, these databases are suitable for fast writes.

Document databases store text, media, and JSON or XML data. The value in a row is a blob of the aforementioned data and can be retrieved using a key. If you want to search through multiple documents for a specific string, a document database should be used.

Graph databases store data entities and connections between them as nodes and edges. They are similar to a network database and can be used to calculate shortest paths, social network analysis, and other parameters.

Figure 4-15 depicts a NoSQL application pattern, where HBASE (which is a columnar data store) is used to store log-file data and then accessed by the front-end application to search for patterns or specific occurrences of certain strings.

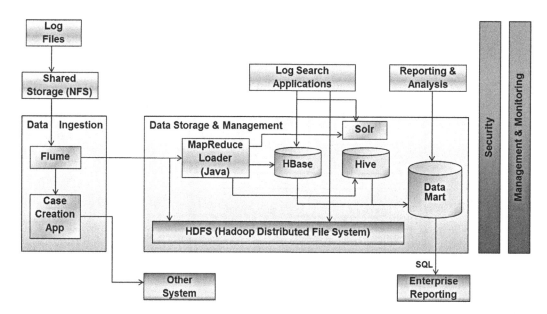

Figure 4-15. *NoSQL pattern—HBase*

Polyglot Pattern

Problem

Can multiple storage mechanisms like RDBMS, Hadoop, and big data appliances co-exist in a solution—a scenario known as "Polyglot Persistence"?

Solution

Certainly. Because the type of data to be stored by an application has changed from being text to other unstructured formats, data can be persisted in multiple sources, like *RDBMS, Content Management Systems (CMS)*, and *Hadoop*. As seen in Figure 4-16, for a single application and for various use cases, the storage mechanism changes from traditional RDBMS to a key-value store to a NoSQL database to a CMS system. This contrasts with the traditional view of storing all application data in one single storage mechanism.

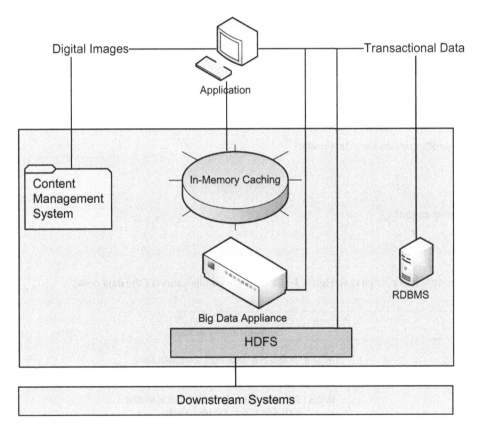

Figure 4-16. *Polyglot pattern*

Big Data Storage Infrastructure

Problem

What role do infrastructure vendors play in the Hadoop implementation?

Solution

As per the IDC (http://www.idc.com/getdoc.jsp?containerId=IDC_P25990), the following are the major vendors:

- **Storage vendors (incumbents):** Dell, EMC, HP, IBM, NetApp, and Oracle
- **Storage vendors (upcoming):** Amplidata, Cleversafe, Compuverde, Coraid, DDN, Nexsan, Nimble, Nimbus, and Violin Memory

Because Hadoop is about distributed storage and analysis, infrastructure vendors play a major role. Vendors like Oracle, EMC and IBM have started packaging infrstructure apart from their big data appliance as part of their Hadoop solutions. The only advantage of such solutions is that the client has to be concerned about only a single vendor. But, again, there is a concern of a being locked to a single vendor and that migrating or decoupling individual entities of a Hadoop ecosystem might become too costly for the client. Due diligence and a total cost of ownership (TCO) assessment needs to be thoroughly done before opting for such a packaged solution.

Typical Data-Node Configuration

Multiple vendors have varying configurations for a data node.

Problem

What is a typical data-node configuration?

Solution

Per information from Intel (www.intel.com/bigdata), Figure 4-17 shows the configuration of the data node.

Entity	Configuration of Data Node
CPU	Two CPU sockets with six or eight cores, Intel Xeon processor E5-2600 series @ 2.9 GHz
Memory	48 GBs (6X8 GBs 1.35v 1333 MHz DIMMs) or 96 GBs (6x16 GBs 1.35v 1333 MHz DIMMs)
Disk	10-12, 1-3 TB SATA drives
Network	1x dual port 10 GbE NIC, or 1x quad port 1 GbE NIC

Figure 4-17. *Intel—Data-node configuration*

Figure 4-18 shows the same information for the IBM Big Data Networked Storage Solution for Hadoop (`http://www.redbooks.ibm.com/redpapers/pdfs/redp5010.pdf`).

Entity	Configuration of Data Node
CPU	Dual Intel Xeon E5-2407 4C/8T 2.2 GHz
Memory	6x8 GBs DDR3 1333 MHz (48 GBs total)
Disk	Sixty 2 TBs 7200 rpm NL- SAS 3.5 disks
	DCS3700 with dual-active Intelligent controllers
Network	Two GbE (1 Gbps) integrated ports
	One 10 GbE 2-port NIC

Figure 4-18. *IBM—Data-node configuration*

Summary

Since multi-structured formats are here to stay, various mechanisms of storage have evolved and are changing the way data storage architecture is designed. As visualization tools take the center stage in the big data world, they will drive how data has to be stored or restructured and necessitate that data be stored in newer formats. But the basic premise of infrastructure capacity planning will still prevail—the only difference being horizontal scaling taking precedence over vertical scaling. Subsequent chapters on data access and data visualization will provide more insight into how the data needs to be stored.

■ ■ ■

CHAPTER 5

Big Data Access Patterns

Traditionally, data was in text format and generally accessed using JDBC adapters from an RDBMS. Unstructured data like documents were accessed from document management systems (DMS) using simple HTTP calls. For performance, improvement concepts like caching were implemented. In the big data world, because the volume of data is too huge (terabytes and upwards), traditional methods can take too long to fetch data. This chapter discusses various patterns that can be used to access data efficiently, improve performance, reduce the development lifecycle, and ensure low-maintenance post-production.

Problem

What are the typical access patterns for the Hadoop platform components to manipulate the data in the Hadoop storage layer?

Solution

Figure 5-1 shows how the platform layer of the big data tech stack communicates with the layers below.

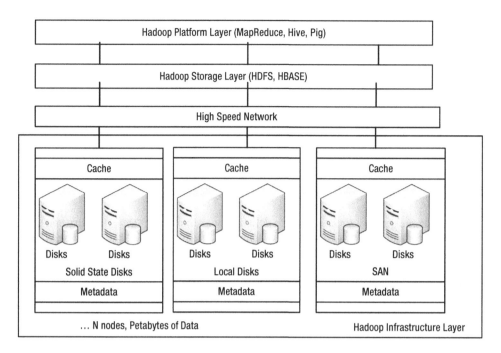

Figure 5-1. *Big data platform architecture*

The *Hadoop platform management layer* accesses data, runs queries, and manages the lower layers using scripting languages like Pig and Hive. Various *data access patterns* (platform layer to storage layer communication) suitable for different application scenarios are implemented based on the performance, scalability, and availability requirements.

Data access patterns describe solutions to commonly encountered problems when accessing data via the storage layer that can be chosen based on performance, scalability, and availability requirements. In the big data world, data that needs to be accessed can be classified as

- Incremental data

- Selective/filtered data

- Near real-time data with low latency

The raw big data does not provide intelligent information about the content and its operation. It is expected that the intended users of the data should be able to apply enough domain knowledge to get any meaningful insight from the raw data.

Data can be accessed from the big data resources in two primary forms:

- **End-to-End User Driven API:** These APIs permit users to write simple queries to produce clipped or aggregated output and throw on a visual display. *Google Search* is an example where the query results are abstracted from the user and the results are fetched using BASE (**b**asically **a**vailable **s**oft state consistent **e**ventually) principles. Google gives users the opportunity to enter a query according to a set of Google-specified query rules, and it provides an output without exposing the internal mechanism of the query processing.

- **Developer API:** Individual developers can interact with the data and analytics service. These services might be available in SaaS (software as a service) formats. Amazon Web Services (AWS) is an example of an API. The API enables querying of the data or the summary of the analytics transactions.

Some of the patterns mentioned in this chapter can be used in conjunction with "data storage patterns." We will cover the following common data access patterns in this chapter, as shown in Figure 5-2:

- **Stage Transform Pattern**: This pattern uses the *end-to-end user API* approach and presents only the aggregated or clipped information in the NoSQL layer (Stage) after transforming the raw data.

- **Connector Pattern:** This pattern uses the *developer API* approach of using APIs for accessing data services provided by appliances.

- **Lightweight Stateless Pattern:** This pattern uses lightweight protocols like REST, HTTP, and others to do stateless queries of data from the big data storage layer.

- **Service Locator Pattern:** A scenario where the different sources of unstructured data are registered on a service catalog and dynamically invoked when required.

- **Near Real-Time Pattern:** This is an access pattern that works well in conjunction with (and is complementary to) the data ingestion pattern "just-in-time transformation."

Understanding Big Data Access

The different big data access patterns are shown in Figure 5-2 and are described in detail in this chapter.

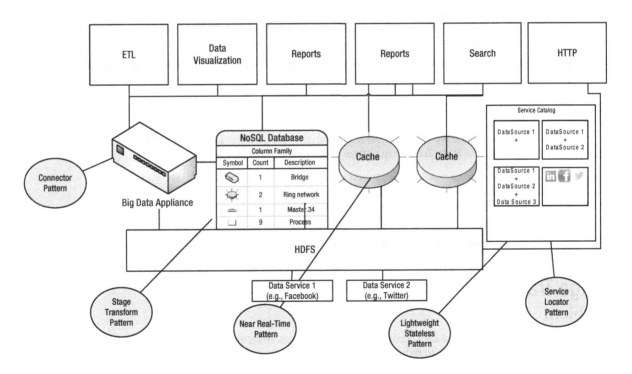

Figure 5-2. *Big data access patterns*

As seen in the discussion of NoSQL in earlier chapters, the way data is stored and structured in a NoSQL database is typically governed by the use cases to be implemented. Along similar lines, patterns for data access are also governed by the application functionality.

As you can see in Table 5-1, data should be abstracted in a layer above the Hadoop distributed file system (HDFS) to ensure low latency and business-specific data storage in a structured format.

Table 5-1. *Use Cases to Access Patterns*

Sr. No	Use Cases	Access Pattern
1	Bulk data	Connector
2	Search	Near Real-Time, Stage Transform
3	Data Visualization	Stage Transform plus Connector
4	Reports	Stage Transform
5	Data Discovery	Lightweight Stateless
6	Enterprise-Wide Dashboard	Service Locator

Different patterns are suitable for different types of use cases:

- The Connector pattern is typically used to process bulk data in XML form. Usually, the connector APIs are provided by appliances or by the business intelligence (BI) systems that are big-data compliant.

- The Stage Transform pattern is useful for rapidly searching data that has been abstracted from HDFS data storage into the NoSQL layer.

- If the data needs to be visualized in different perspectives, the Stage Transform and Connector patterns can be used in conjunction to present the data in different views.

- Standard enterprise reports can be derived from the NoSQL databases instead of HBASE directly using the Stage Transform pattern.

- Data discovery from multiple data sources can be facilitated using RESTful services provided by those sources using the Lightweight Stateless pattern.

- An enterprise-wide dashboard collates data from applications across the organization using the catalog of services available from the API management software. This dashboard can dynamically present the data using the Service Locator pattern.

Stage Transform Pattern
Problem

HDFS does not provide the ease of data access that an RDBMS does. Also, there is too much data that is not relevant for all business cases. Is there a way to reduce a huge data scan?

Solution

HDFS is good for two purposes:

- Data storage

- Data analytics

As mentioned earlier, NoSQL does not need to host all the data. HDFS can hold all the raw data and only business-specific data can be abstracted in a NoSQL database, with HBase being the most well-known. There are other NoSQL databases—like *MongoDB*, *Riak*, *Vertica*, *neo4j*, *CouchDB*, and *Redis*—that provide application-oriented structures, thereby making it easier to access data in the required format.

For example, for implementing data discovery for a retail application that depends on social media data, enterprise data, historical data and recommendation engine analysis, or abstracting data for a retail user or users, a NoSQL database makes the implementation of a recommendation engine much easier.

The stage transform pattern in Figure 5-3 can be merged with the NoSQL pattern, which was discussed in Chapter 4 of. The NoSQL pattern can be used to extract user data and store it in a NoSQL database. This extracted data, which will be used by the recommendation engine, significantly reduces the overall amount of data to be scanned. The performance benefit recognized will invariably improve the customer experience.

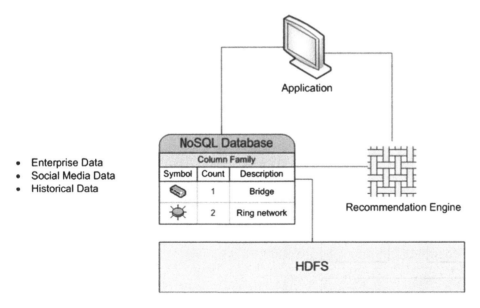

Figure 5-3. *Stage Transform pattern*

As you can see in Figure 5-3, the two "stages" of HDFS and NoSQL storage are used appropriately to reduce access times. Frequently accessed information is aggregated or contextualized in the NoSQL layer. The HDFS layer data can be scanned by long-running batch processes to derive inferences across long periods of time.

This virtualization of data from HDFS to a NoSQL database is implemented very widely and, at times, is integrated with a big data appliance to accelerate data access or transfer to other systems, as can be seen in the section that follows.

Connector Pattern

Problem

Just as there are XML accelerator appliances (like IBM DataPower), are there appliances that can accelerate data access/transfer and enable the use of the developer API approach?

Solution

EMC Greenplum, IBM PureData (Big Insights + Netezza), HP Vertica, and *Oracle Exadata* are some of the appliances that bring significant performance benefits. Though the data is stored in HDFS, some of these appliances abstract data in NoSQL databases. Some vendors have their own implementation of a file system (such as GreenPlum's OneFS) to improve data access.

The advantage of such appliances is that they provide *developer-usable APIs* and SQL-like query languages to access data. This dramatically reduces the development time and does away with the need for identifying resources with niche skills.

Figure 5-4 shows the components of a typical big data appliance. It houses a complete big data ecosystem. Appliances support virtualization. Thus, each node/disk is a virtual machine (VM) on top of a distributed database like HDFS. The appliance supports redundancy and replication using protocols like RAID. Some appliances also host a NoSQL database.

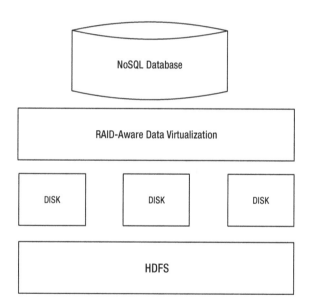

Figure 5-4. *Big data appliance typical configuration*

Examples are shown in Table 5-2.

Table 5-2. *Big Data Appliance Products*

Vendor	Remarks
Aster Teradata	35% faster CPU, 5 times more memory, 20 times faster throughput, 60% smaller datacenter footprint compared to other vendors
EMC Isilon OneFS	Multiple Isilon appliances serve different purposes: • IOPS-intensive application appliance • High-concurrency and throughput-driven workflows appliance • Near-primary accessibility, with near-tape value appliance • Performance Accelerator appliance • Backup Accelerator appliance

Appliances might induce dependency on vendors. Some appliances, as seen in an earlier chapter, come packaged as hardware, software, or a NoSQL database. For example, Vertica comes bundled with built-in "R" and "SAS" based engines and algorithms. Vertica can support any Hadoop distribution, such as *Hortonworks*, *Cloudera*, and *MapR*.

Near Real-Time Access Pattern

When we talk about "near real-time" access, we should keep in mind two things:

- Extremely low latency in capturing and processing the data. This means that as events happen, you act on the data; otherwise, that data becomes meaningless in the next minute.

- Analyzing the data in real time. This means you will need to have sophisticated analysis patterns to quickly look at the data, spot anomalies, relate the anomalies to meaningful business events, visualize the data, and provide alerts or guidance to the users. All this needs to happen at that very moment.

While the Hadoop ecosystem provides you the platform to access and process the data, fundamentally it still remains a batch-oriented architecture.

In this context, we encounter technologies used by Storm, in-memory appliances like Terracota, heavily indexed search patterns through Lucene and Solr.

Problem

Can we access data in near real-time from HDFS?

Solution

Near real-time data access can be achieved when *ingestion, storage, and data access* are considered seamlessly as one single "pipe." The right tools need to be used to ingest, and at the same time data should be filtered/sorted in multiple storage destinations (as you saw in the multidestination pattern in an earlier chapter). In this scenario, one of the destinations could be a cache, which is then segregated based upon the business case. That cache can be in the form of a NoSQL database, or it can be in the form of memcache or any other implementation.

A typical example is searching application logs where data for the last hour is needed.

As you can see in Figure 5-5, the moment the data is ingested and filtered, it is transferred to a cache. This is where 90% of the noise is separated from 10% of the really relevant information. The relevant information is then stored in a rapidly accessible cache, which is usually in-memory. To quickly analyze this information before it becomes stale, search engines like Solr are used to complete this "Near Real-Time Access pattern" scenario.

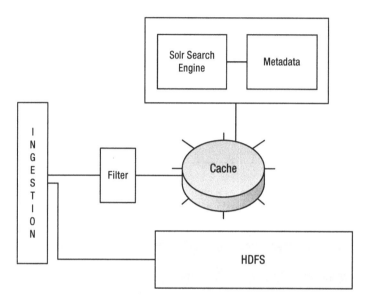

Figure 5-5. *Near Real-Time Access pattern using multicache*

Lightweight Stateless Pattern
Problem

NAS (Network Access Storage) provides single file access. Can HDFS provide something similar using a lightweight protocol?

Solution

Files in HDFS can be accessed over RESTful HTTP calls using WebHDFS. Since it is a web service, the implementation is not limited to Java or any particular language. For a cloud provider or an application wanting to expose its data to other systems, this is the simplest pattern.

The Lightweight Stateless pattern shown in Figure 5-6 is based on the HTTP REST protocol. HDFS systems expose RESTful web services to the consumers who want to analyze the big data. More and more of these services are hosted in a public cloud environment. This is also the beginning of the Integration Platform as a Service (iPaaS). This pattern reduces the cost of ownership for the enterprise by promising a pay-as-you-go model of big data analysis.

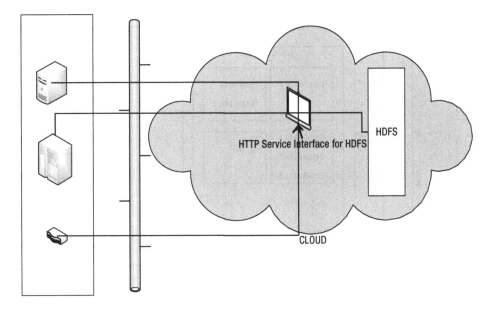

Figure 5-6. *Lightweight Stateless pattern for HDFS*

Service Locator Pattern

Problem

If there are multiple data storage sites (for example, Polyglot persistence) in the enterprise, how do I select a specific storage type?

Solution

For a storage landscape with different storage types, a data analyst needs the flexibility to manipulate , filter, select, and co-relate different data formats. Different data adapters should also be available at the click of a button through a common catalog of services. The Service Locator (SL) pattern resolves this problem where data storage access is available in a SaaS model.

Figure 5-7 depicts the Service Locator pattern. Different data sources are exposed as services on a service catalog that is available to data analysts based on their authorization. The services could be within the enterprise or outside of it. Different visualization tools can mix and match these services dynamically to show enterprise data alongside social media data.

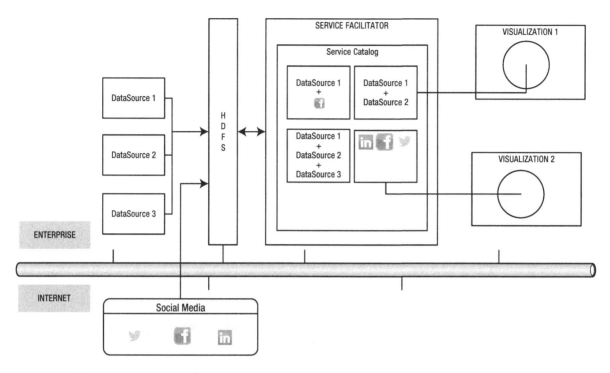

Figure 5-7. *Service Locator pattern for HDFS*

Rapid Data Analysis

Problem

Is MapReduce the only option for faster data processing and access?

Solution

No. There are alternatives like Spark and Nokia's DISCO.

Spark is an open source, cluster-computing framework that can outperform Hadoop by 30 times. Spark can work with files stored in HDFS. MapReduce relies on disk storage while Spark relies on in-memory data across machines.

Figure 5-8 shows a comparison of the performance of a Spark vs. MapReduce.

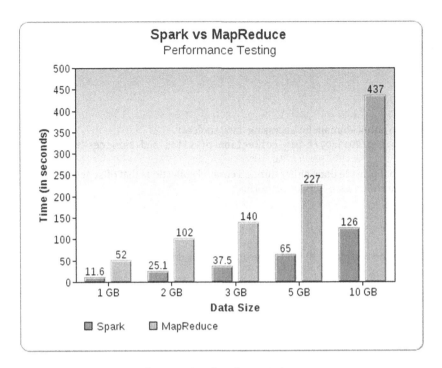

Figure 5-8. *Comparison between Spark and MapReduce*

Secure Data Access

Problem

What security measures can be included to ensure data is not compromised during the interlayer communication?

Solution

Typical security measures that need to be looked into include the following:

- Confidentiality: Data should be encrypted so that it is not sniffed during transport between the layers.

- Authentication: Only authenticated users should be given access.

- Authorization: Users should have access to data according to their profiles and access rights only.

Other security measures are the traditional data center security measures like these:

- Network Intrusion Detection Systems (NIDS)

- Providing access only to requests coming from a particular IP

- Running nodes on ports other than default ports

- Host-based intrusion-prevention systems

Problem

Are there any large datasets available in the public domain that can be accessed by a layperson to analyze and use for big data experimentation?

Solution

Yes, there are many sites and services in the public domain for accessing data, such as:
`http://www.visualisingdata.com/index.php/2013/07/a-big-collection-of-sites-and-services-for-accessing-data/`.

This collection presents the key sites that provide data, either through curated collections that offer access under the open data movement or through software/data-as-a-Service platforms.

Problem

Are there products from industry leaders in the traditional BI landscape that offer big data integration features?

Solution

Yes, vendors like *Pentaho*, *Talend*, *Teradata*, and others have product offerings that require less learning time for BI developers to harness the power of big data.

Example: Pentaho's big data analytics integrates with Hadoop, NoSQL, and other big data appliances. It's a visually easy tool that can be used by business analysts.

Summary

Big data access presents a unique set of issues that can be addressed using the set of patterns described in this chapter. As big data access becomes more secure and frameworks like MapReduce evolve (for example, YARN), newer data storage and access patterns will emerge. The need to access data in real time requires the usage of techniques like memcache, indexing, and others. This area is evolving, and many new research projects are underway that will lead to new patterns of usage. The key takeaway for big data architect is to note that the access patterns have to be used in conjunction with the right data-storage pattern to ensure the best performance and lowest latency.

CHAPTER 6

■ ■ ■

Data Discovery and Analysis Patterns

Big data analysis is different from traditional analysis as it involves a lot of unstructured, non RDBMS types of data. This type of analysis is usually related to text analytics, natural language processing. Areas like video and image analytics are still evolving. Big data analysis attempts to interpret and find insightful patterns in the customer behavior that perhaps the sales force already had some idea about, but did not have the data to support it. Big data analysis methods are used to analyze social media interactions, bank transactions for fraud patterns, customer sentiments for online product purchases, etc. Let's look at some patterns that may help discover and analyze this unstructured data.

Problem

What are the different types of unstructured data sources that are analyzed in a big data environment?

Solution

There are different types of unstructured data hidden in multiple data sources that are available as large datasets:-

- Documents contain textual patterns, repetitions of certain words, etc. that can be analyzed and interpreted.

- Application logs contain a wealth of information about upcoming down time, applications that are coming up for maintenance and upgrade, etc.

- E-mail has become the defacto means of communication both in corporate as well as informal channels.

- Social media forums like Yammer, Twitter, and Facebook generate a lot of text and symbols may that determine customer behavior.

- Machine generated data like RFID feeds, weather data, etc. also provide a large data set for automated analysis.

Problem

What are different statistical and numerical methods available for analyzing the different unstructured data sources?

Solution

Various methods that have their origins in computer science computational methods exist for analyzing big data sources:-

- Natural language processing

- Text mining

- Linguistic computation

- Machine learning

- Search and sort algorithms

- Syntax and lexical analysis

Using these methods, the output of the analysis of the results is combined with the structured data to arrive at meaningful insights.

Problem

What are the typical analysis patterns used for analyzing big unstructured data?

Solution

We will cover the following analysis patterns (shown in Figure 6-1) in this chapter:-

- **Data Queuing Pattern:** This pattern is used to handle spikes in the data being analyzed. A lightweight process or workflow is required to queue the additional chunks of data and then route them to available nodes.

- **Index-based Insight Pattern:** This is a data discovery pattern in which a series of indexes are defined based on inputs from users who interact with customers. These indexes are tuned iteratively as more and more data determines the range of the indices.

- **Constellation Search Pattern:** This pattern utilizes master data management (MDM) concepts where a constellation of metadata is used to confirm the repetitive occurrence of a set of variables. This constellation is refined and the added back to the MDM system.

- **Machine Learning Pattern:** Statistical and numerical analysis algorithms are applied using programming tools to identify patterns in machine generated data from energy meters, weather related devices, RFID feeds, etc.

- **Converger Pattern:** Analyzing unstructured data and then merging it with structured data is required to get the enterprise wide perspective to make decisions.

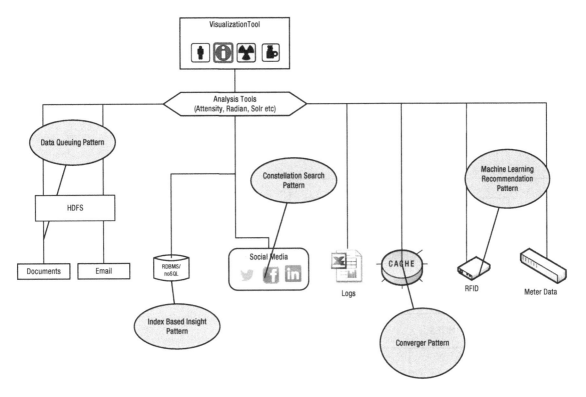

Figure 6-1. *Analysis Patterns*

Data Queuing Pattern

Problem

Unlike traditional structured data where the volume of data is known and predefined, unstructured data comes in spurts. The big data analysis tools may be analyzing different volumes of data. How do I manage the changing volume spikes?

Solution

Events like professional football or rock concerts trigger a lot of activity in different forums and email groups. It is also the right time to roll out offers and promotions.

To handle such spikes in data we can use cloud infrastructure as a service (IaaS) solutions.

A simple lightweight workflow to queue the additional data chunks and orchestrate the assignment of analysis to the nodes that are free is required in the architecture. There is also a need for spining new virtual machines, on demand, to address the new capacity requirements dynamically.

The data queuer (shown in Figure 6-2) that sits above the HDFS layer allows us to provision and orchestrate the analysis payload so that it does not interrupt the analysis tools and provides a seamless interface.

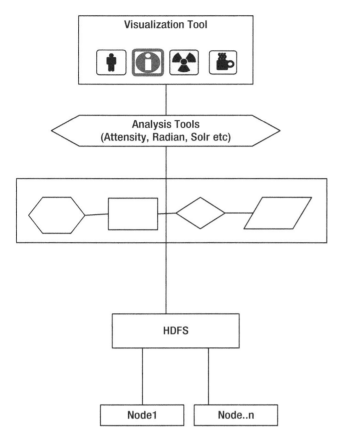

Figure 6-2. *Data Queuing Pattern*

Index based Insight Pattern

Problem

I need to estimate the number of different types of parameters that I am monitoring, e.g., all parents who buy toys, all children above 13 in a neighborhood, etc. This is a count that needs to be averaged out to reach a stable count. How do I setup an analysis pattern that helps me to index such variables and provide insight?

Solution

The above problem requires an efficient key / index lookup that provides rapid scanning and also helps to keep related column families together. This is a pattern used by many analysis tools to build indexes and enable rapid search. Indexes can be used along with zones and/or partitions to improve performance of 'read' requests.

As data grows and read requests vary, more indexes need to be incorporated based on the most frequently 'read' data attributes.

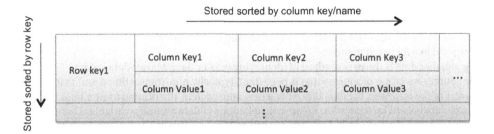

Figure 6-3. *Columnar index based Insight Pattern*

In fact, this pattern has been extended further by graph databases where the attributes and relationships among 'Nodes' is dynamically added as queries grow more complex.

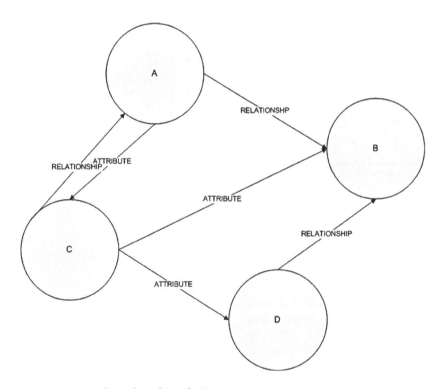

Figure 6-4. *Attributes based Insight Pattern*

Constellation Search Pattern
Problem

How do I spot groups of related data and define metadata when the database is not designed for this type of analysis upfront before the data is loaded as in a traditional RDBMS?

Solution

Some of the common big data social media analytics use cases help in identifying groups of people with the proper attributes who can then be targeted in a focussed manner.

The criteria for grouping together data is the new method of identifying master data. Master data is data that is generally from a single source or a group of sources that is persistent, typically non-transactional and that is important in analyzing aspects of the business. Big data is driving a new approach to distributed master data management (D-MDM). Master data is used by analysts to create constellations of data around a variable. There can be constellations at various levels of abstraction to give different views from different "levels".

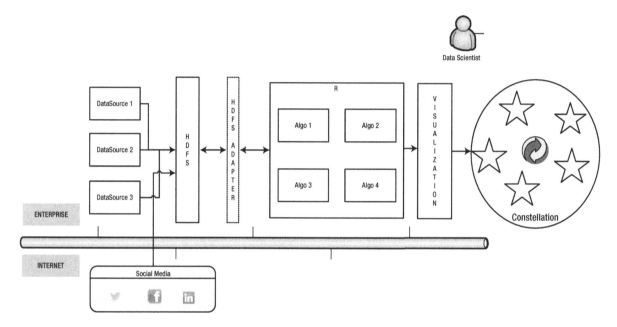

Figure 6-5. *Constellation Search pattern*

As seen in Figure 6-6, the data, typically stored in a Hadoop cluster is a combination of data from social media as well as the data in the enterprise. The information from the social media data can be analyzed to extend the metadata stored in D-MDM. This extended meta-data, along with the social-media analytics, can be used to create a constellation of information that can be used to arrive at newer insights.

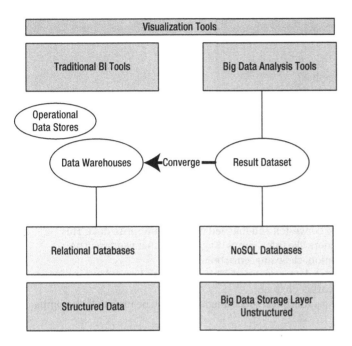

Figure 6-6. *Coverger Pattern*

For example, Facebook's 'Needle in a haystack' architecture uses a combination of in-memory indexes and metadata to search photos and 'photos related to a photo'. The index stores the minimal metadata required to locate a particular needle in the haystack. This efficient usage of index and metadata combinations not only helps reduce search latency but also provides a constellation of relationship meta-data around an image. This constellation can then be used for performing further anlaysis and arriving at meaningful and hitherto unknown insights.

Machine Learning Recommendation Pattern
Problem

What analysis pattern should be used for meter data, RFID feeds, etc?

Solution

Machine learning is about turning data into information using automated statistical methods without direct human intervention. With the deluge of data spewed out by various devices it is not humanly possible to define analysis patterns. We have to rely on statistical methods and algorithms to do this task for us.

The various machine learning algorithms usually used to analyze big data are:-

- kNN Classification algorithm to identify your neighbors

- Time decision trees

- Bayesian decision theory

- Regression coefficients

- Apriori algorithms for finding frequently occurring data items

Converger Pattern

Problem

How do I analyze both traditional and external data together?

Solution

Social media (Facebook, LinkedIn, Twitter, Google+ and others) carries views and opinions which can be merged with the analysis done using organizational data to bring more insight into 'sentiment' regarding an organization or a product. This combined analysis can help with insights that the data present within the enterprise system cannot. To achieve this, the architecture should support combining the external analysis with the in-house analysis results.

As seen earlier in the question on the 'Constellation Search' pattern, the data from social media and the data within the enterprise needs to be co-related using D-MDM to arrive at meaningful insights. For this to happen, the format of the data from external sources needs to be converted and married with the enterprise data. This convergence can be arrived at by using data from providers like InfoChimps, Kaggle and other vendors. The convergence involves typical ETL steps like, transformation, cleansing, enrichment et al.

The convergence has to happen before the enterprise data is analyzed. This pattern can be used in consonance with the 'Facade Data storage' pattern discussed in an earlier chapter.

The constellation arrived at using machine learning patterns on social-media data can be used to look at impact on revenues, brand image, churn rates, etc.

Challenges in Big Data Analysis

Problem

What are the most prevalent challenges in big data analysis?

Solution

Big data analysis has new challenges due to its huge volume, velocity, and variety. The main among them are:-

- **Disparate and insufficient data:** Plain text in any language is different from person to person communications. In the case of big data, traditional analytics algorithms fail due to the heterogeneity of the data. A very effective data cleansing process is required to tackle this challenge before it is fit for analysis.

- **Changing volumes:** Technologies like Hadoop allow us to manage the large volumes at relatively low cost, however the frequency of these change in volumes can impact the performance of analysis. Use of the cloud infrastructure combined with the data queuing pattern can help manage this challenge.

- **Confidentiality:** The confidentiality of the data being analyzed which belongs most often to individual users and customers make the handling of big data a very debatable issue. Laws are being framed to protect the privacy rights of while their data is being analyzed. The analysis methods have to ensure that there is no human intervention and inspection of the data. Machine learning becomes very important in this context.

- **Performance:** With options like data analysis as a service (DaaS) available for outsourcing your big data analysis, performance and latency become a big challenge. Ingenious ways of streaming data have to be available for fast transfer of data to overcome this challenge.

Frameworks like AppFabric and Open Chorus help solve some of the above challenges.

Log File Analysis

Problem

Log files are the most unstructured type of data. They are defined differently by each product as well as by individual. Is there a high level approach to tackle this problem?

Solution

Log files are generated by all applications and servers and are defined by users, developers, and system administrators.

Operating systems and application servers generate huge amounts of data into log files. Big data analysis tools have to cleanse, decipher, analyze, and mine relevant information from these strings of raw data. Analyzing this information will give information pertaining to the health of these systems. Organizations can create a service catalogue of private cloud offerings based on the usage statistics of servers and infrastructure.

Most application and web servers allow developers and administrators to define some sort of loose format for the logs, generally using a separator symbol between attributes. The first step to start analyzing these files is to understand these formats and create a data model for example, map each block of a string to attributes like Http 404 codes or events that have a purchase action, etc.

Tools like Splunk provide a methodical approach to analyzing log information. The typical steps in analyzing log information are:-

1. Identify attributes being logged in the log

2. Make note of exceptions

3. Create a data mapping of attributes

4. Scan the raw data repeatedly to detect frequency of events and exceptions

5. Select the time period that you want to select the data asset for

6. Find categories for repetitive events and correlate

7. Fine tune the categories with every search

8. Run analytics on the set of attributes and categories that have been stabilized after a few iterations.

Sentiment Analysis

Problem

Organizations are making instant decisions based on the sentiments, opinions and views of their customers. These opinions are present in tweets, blogs and forums. How do I analyze this information?

Solution

Sentiment analysis involves analyzing social media text for people's opinions. As you can imagine it involves understanding language idiosyncrasies as well. Lexical analysis can be applied to formal and structured documents as they are usually grammatically correct. However blogs and tweets use slang, that is difficult to analyze and ambiguous. The outcome of a sentimental analysis is usually a percentage or strength range that is determined based on the frequency of occurrence of words, parts of speech used, use of negation syntax and position of the words in a sentence. Techniques like Bayes' theorem, entropy and vector analysis are used to infer from textual data. Tools like Splunk are used to facilitate sentiment analysis.

Data Analysis as a Service (DaaS)

Problem

Can I outsource my big data analysis to a third party vendor?

Solution

Yes. APIs provided by many vendors like the Google Prediction API provide the ability to analyze your data without having to invest in a large capital expense to build your big data analysis platform. The APIs are usually *Restful* and involve just invocation of the service over HTTP. The data interchange is also lightweight and loosely structured using JSON objects or some other data format.

Summary

Analysis tools have been upgraded to analyze big data residing on Hadoop clusters, in-memory or in social media networks. Patterns are required to obtain the same performance, confidentiality and context from the traditional analytics tools using statistical analysis, cloud services and sentiment analysis to map to the traditional reporting means like dashboards and reports.

With these patterns you have seen how the data scientist can be augmented by the architect's solutions, to supplement his data mining repertoire of algorithms and methods.

CHAPTER 7

■ ■ ■

Big Data Visualization Patterns

Pie charts and bar charts have been the most common and often-used analytical charts. Data interpretation, per se, has now moved from a representation of sample data to a full-fledged analysis of all the data residing within the enterprise and also the "sentiments" that the data from social media analytics churns out. This chapter introduces commercial visualization tools in the big data domain, the use-cases that utilize the power of the Internet to derive meaningful insights, the marriage of machine-learning and visualization, and observations about how it is changing the landscape for business analysts and data scientists.

Introduction to Big Visualization
Problem

How is big data analysis different from the traditional business intelligence (BI) reporting?

Solution

Analysis of data has not always necessarily been conducted on all the data residing within the enterprise. (See Table 7-1.) It is sometimes sample data culled for a limited time period, or it is the volume of data that the reporting tools could handle to create a time-bound time report.

Table 7-1. *Traditional BI vs. Big Data*

Traditional BI Reporting	Big Data Analysis
Reporting tool like Cognos	Visualization tool like QlikView or Tableau
Sample data	Huge volume of data
Data from the enterprise	Data from external sources like social media apart from enterprise data
Based on statistics	Based on statistics and social sentiment analysis or other data sources
Data warehouse and data mart	OLTP, real-time as well as offline data

There was a need to break the mold of restricting reports to pie and bar charts and also to run reports on the full volume of data at the disposal of the enterprise. And that had to be done within the boundaries of time limits or maintenance windows. This necessitated the following:

- Storage for large volumes of data

- Business-specific visualization

- Faster processing of large volumes of data

- Ease of tool usage for data scientists and business analysts

As depicted in Figure 7-1, traditional reporting follows a sequential process of transferring OLTP data to a data warehouse, running statistical and analytical algorithms on the de-normalized data, and churning out reports in patterns like bar graphs, and pie charts, and others.

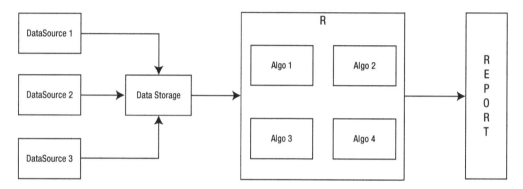

Figure 7-1. *Traditional visualization*

Big Data Analysis Patterns

Problem

What are the new big data analysis and visualization patterns that enable you to gain more insight from the huge volume of data?

Solution

Traditional analysis and visualization techniques need to be modified to provide that "helicopter" view of a large volume of data. The patterns will not be visible if the data is viewed in very granular detail. Visualization tools and graphics have to follow a more "planetary" view, where the data scientist is like an astronomer trying to find a new star or black hole in a huge distant galaxy.

Some of the analysis patterns mentioned in this chapter can be used in conjunction with "data access patterns." We will cover the following common data-analysis patterns in this chapter as shown in Figure 7-2:

- **Mashup View Pattern:** This pattern is used to maximize the performance of the queries by storing an aggregated mashup view in the HIVE layer that functions as a data warehouse. The MapReduce jobs are run in batches to update the warehouse offline.

- **Compression Pattern:** This pattern compresses, transforms, and formats data in a form that is more rapidly accessible.

- **Zoning Pattern:** Data can be split and indexed based on various attributes in different zones for faster access.

- **First Glimpse Pattern:** A scenario where the visualization is minimalist and provides a First Glimpse of the most relevant insights. A user can pull more information if required, which can be fetched in the interim while he is viewing the First Glimpse.

- **Exploder Pattern:** An extension of the First Glimpse pattern that allows you to visualize data from different sources in different visual perspectives.

- **Portal Pattern:** An organization that has an existing enterprise portal can follow this pattern to re-use the portal for visualization of big data.

- **Service Facilitator Pattern:** A pay-as-you go approach to big data analysis projects.

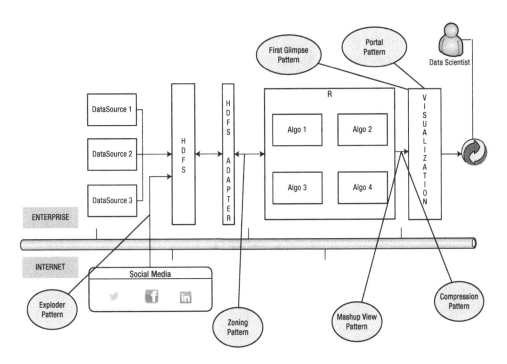

Figure 7-2. *Big data analysis and visualization patterns*

Problem

How do you overcome the limitations of existing reporting tools?

Solution

Commercial tools have emerged in the market that promise higher throughput over a large volume of data and provide business-specific visualizations. Here are some of the commercially known tools:

- QlikView

- TIBCO Spotfire

- SAS RA

- Tableau

These tools, along with market-known machine-learning tools (based on the R language) like "Revolution R" from Revolution Analytics, can be a good combination to ensure meaningful data discovery and visualization.

Mashup View Pattern
Problem

It takes a very long time to analyze data using MapReduce jobs. Is there a way to improve the performance?

Solution

HIVEover Hadoop, though good at storage and at running MapReduce jobs, is unable to do a good job when running complex queries consisting of JOINs and AGGREGATE functions.

Though most visualization and analytical tools can talk to Hadoop via HIVE queries, as in traditional methods, it makes sense to create an aggregated mashup view either within Hadoop or in abstracted storage like RDBMS/NoSQLl/Cache as shown in Figure 7-3. The Mashup View pattern reduces analysis time by aggregating the results of the MapReduce queries in the HIVE data warehouse layer.

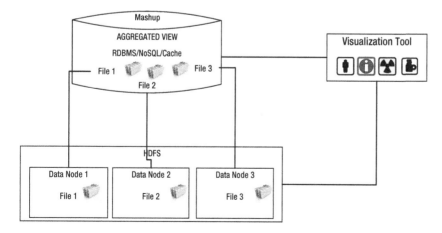

Figure 7-3. *Mashup View pattern with data abstracted from Hadoop*

As shown in Figure 7-4, the mashup can be achieved within the Hadoop layer also, instead of the HIVE layer, to save expensive storage dollars.

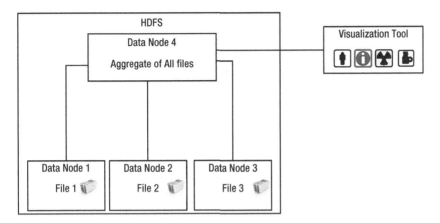

Figure 7-4. *Mashup View pattern with data abstracted within Hadoop*

This strategy is endorsed by many vendors and is provided by the following products in the market:

- IBM Netezza
- Cassandra
- HP Vertica
- Cloudera Impala
- EMC HAWQ
- Hortonworks Stinger

These products provide performance/latency benefits because they access the storage via aggregated views stored in HIVE or in the Hadoop layers, which play the role of a data warehouse.

Compression Pattern
Problem

Is there a faster way to access data without aggregating or mashing up?

Solution

Analysis tools like R support different compression formats—for example, .xdf (*eXtended Data Format*)—as shown in Figure 7-5. Instead of data being fetched from data storage, it can be converted to formats that R understands. This transformation not only provides performance benefits, but also ensures that data is valid and can be checked for correctness and consistency.

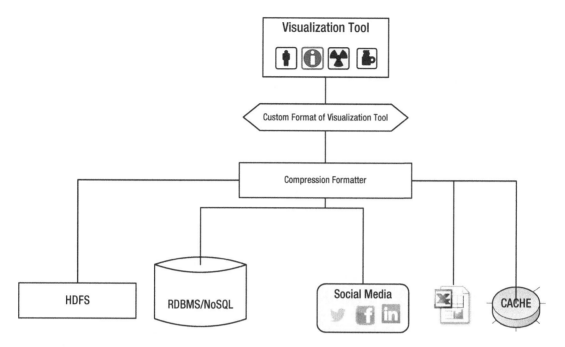

Figure 7-5. *Compression pattern*

Zoning Pattern

Problem

Can I divide and rule the data in a fashion that is characterized by the attributes and hence is easier to locate?

Solution

As shown in Figure 7-6 data can be partitioned into zones at each layer (Hadoop, abstracted storage, cache, visualization cluster) to ensure that only the necessary data is scanned. The data can be divided (or partitioned) based on multiple attributes related to the business scenario in question.

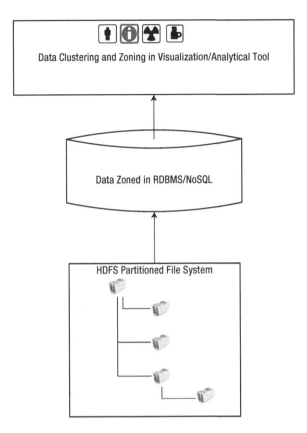

Figure 7-6. *Zoning pattern*

First Glimpse Pattern

Problem

Do I need to see all the results always in a single view?

Solution

Since the volume of data is too huge, it makes sense to fetch only the amount of data that is absolutely essential and provide only the "first glimpse." The *First Glimpse* (FG) pattern shown in Figure 7-7 recommends what is popularly known as "lazy-loading." Let the end user decide how deep he/she wants to drill-down into details. Drill-down data should be fetched only if the user navigates into the subsequent layers of detail.

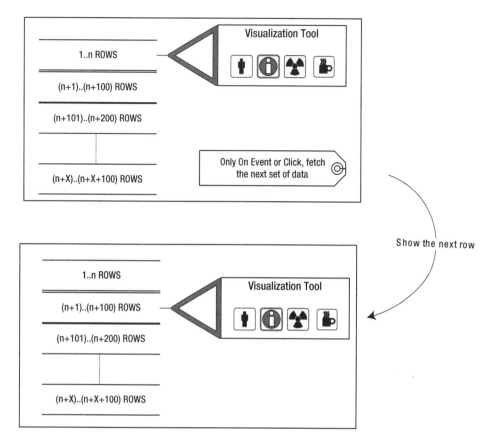

Figure 7-7. *First Glimpse pattern*

Exploder Pattern
Problem

Do I need to see all the results always in a single view and be restricted to a similar visual pattern for the entire data?

Solution

This is an extension of the *First Glimpse* pattern. As shown in Figure 7-8, the difference is that the data may be fetched from a different source or might explode into an altogether different indexed data set. Also, the drill down on a click may produce a different chart type or visualization pattern.

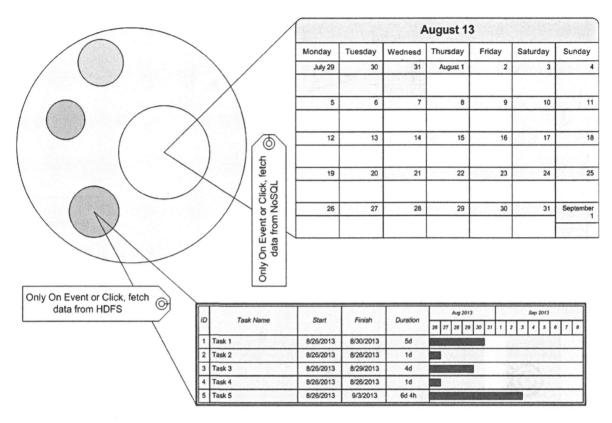

Figure 7-8. *Exploder pattern*

This pattern allows the user to look at different data sets, co-relate them, and also look at them from different perspectives visually.

Portal Pattern

Problem

I have already invested in an enterprise portal. Do I still need a new visualization tool?

Solution

If an organization is already using a web-based reporting solution and wants to continue without introducing new tools, the same existing portal can be enhanced to have a new frame with scripting frameworks like D3.js to enhance the legacy visualization. As shown in Figure 7-9, this ensures that the enterprise does not have to spend money on a new visualization tool.

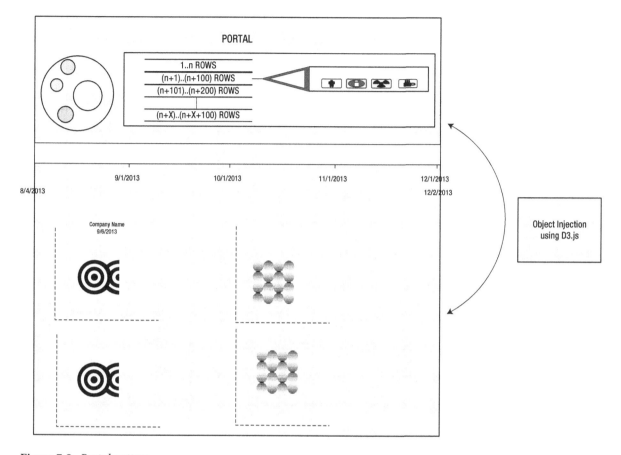

Figure 7-9. *Portal pattern*

Service Facilitator Pattern
Problem

I do not want to invest in the software and hardware to do my first pilot projects. Is big data analysis service available on a pay-as-you-go basis?

Solution

Big data analysis capabilities are available in an *"as-a-Service mode"* using cloud-enabled services—for example,

- Analytics as a Service
- Big Data Platform as a Service
- Data Set Providers

Organizations can look upon these providers as an alternative to circumvent infrastructure and/or skill constraints. Table 7-2 shows a list of "as-a-service" services available and providers.

Table 7-2. *Providers for Various "as-a-service" Services*

Service	Service Provider
Cloud Provider	Amazon Web Services, Infochimps, Rackspace
Platform Provider	Amazon Web Services, IBM
Data Set Provider	UNICEF, WHO, World Bank, Amazon Web Services
Social Media Data Provider	Infochimps, Radian 6

These services can be searched on a service catalog and then used on a pay-as-you-go basis.

As shown in the Figure 7-10 dataset, providers provide large datasets that can complement an organization's existing business cases. The data can be from a government organization, an NGO, or an educational institute, which can be leveraged to complement the analytical data coming from implementations of *Google Analytics* or *Omniture*.

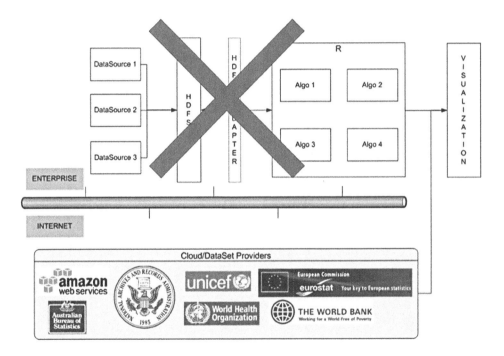

Figure 7-10. *Service Facilitator pattern*

Summary

Traditional reporting tools have been replaced by analytical and visualization tools that can read data from a Hadoop cluster, analyze data in-memory, and display it in a format that the business understands and in a more visually appealing form compared to traditional pie and bar charts. The business intelligence layer is now equipped with advanced big data analytics tools, in-database statistical analysis, and advanced visualization tools, in addition to the traditional components such as reports, dashboards, and queries.

With this architecture, business users see the traditional transaction data and big data in a consolidated single view. Business analysts and data scientists can now look beyond Excel sheets and reporting tools to create visually alluring graphs that can cater to huge volumes of data.

CHAPTER 8

■ ■ ■

Big Data Deployment Patterns

Big data deployment involves distributed computing, multiple clusters, networks, and firewalls. The infrastructure involves complicated horizontal scaling, and the inclusion of the cloud in some scenarios makes it more complex. This chapter illustrates deployment patterns you can use to deal with this complexity up front and align with the other patterns across various layers.

Big Data Infrastructure: Hybrid Architecture Patterns

Infrastructure for a big data implementation includes storage, network, and processing power units. In addition to Hadoop clusters, security infrastructure for data traffic from multiple data centers and infrastructure for uploading data to downstream systems and/or a data center might be needed. Appliances or NoSQL data storage layers might require additional infrastructure for storage of data and metadata.

    A variety of products and/or services can be used to implement a hybrid infrastructure. Various hybrid architecture patterns are discussed in the next sections. Each pattern is presented as a problem in the form of a question, followed by an answer in the form of a diagram.

Traditional Tree Network Pattern
Problem

Ingesting data into a Hadoop platform sequentially would take a huge amount of time. What infrastructure pattern can help you ingest data as fast as possible into as many data nodes as possible?

Solution

Implement a *traditional tree network pattern* (Figure 8-1).

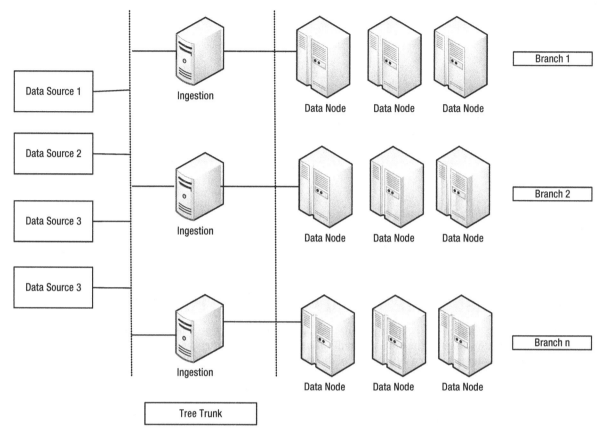

Figure 8-1. *Traditional tree network pattern*

Uploading or transferring bulk data into the Hadoop layer is the first requirement encountered in finalizing the Hadoop infrastructure. As discussed in Chapter 3, Flume or SFTP can be used as the ingestion tool or framework, but until the data is uploaded, typically in terabyte-scale volumes, into the Hadoop ecosystem, no processing can start. Sequential ingestion would consume hours or days. To reduce the ingestion time significantly, simultaneous ingestion can be effected by implementing the traditional tree network pattern. This pattern entails using Flume or some alternative framework to channelize multiple agents in multiple nodes (trunks) that run in parallel and feed into Hadoop ecosystem branches.

Resource Negotiator Pattern for Security and Data Integrity
Problem

When data is being distributed across multiple nodes, how do you deploy and store client data securely?

Solution

Implement a *resource negotiator pattern* (Figure 8-2).

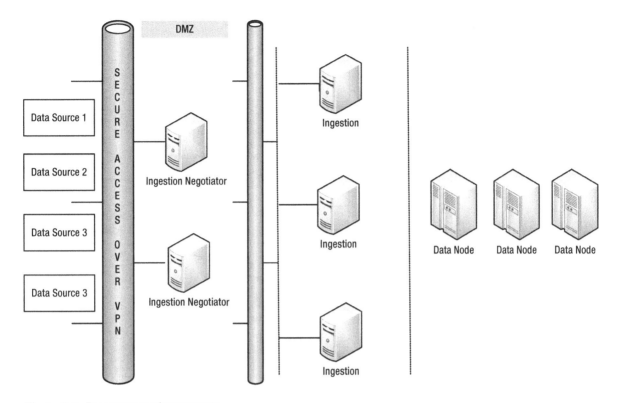

Figure 8-2. *Resource negotiator pattern*

Clients might be wary of transferring data to the Hadoop ecosystem if it is not secure or if the channel through which the data is uploaded is not secure. For its part, the data center in which the Hadoop ecosytem resides might not want to expose the Hadoop cluster directly, preferring to interpose a proxy to intercept the data and then ingest it in the Hadoop ecosystem.

Proxy interposition is effected by implementing a *resource negotiator pattern* (Figure 8-2). Data from the client source is securely ingested into *negotiator nodes*, which sit in a different network ring-fenced by firewalls. Discrete batch job flows ingest data from these negotiator nodes into the Hadoop ecosystem. This double separation ensures security for both the source data center and the target Hadoop data center.

The high processing, storage, and server costs entailed by negotiator nodes might urge you to replace them with an intermediate Hadoop storage ecosystem, as depicted in Figure 8-3. This latter solution, which is particularly appropriate for configurations in which data is being ingested from multiple data centers, gives rise to *spine fabric* and *federation patterns*.

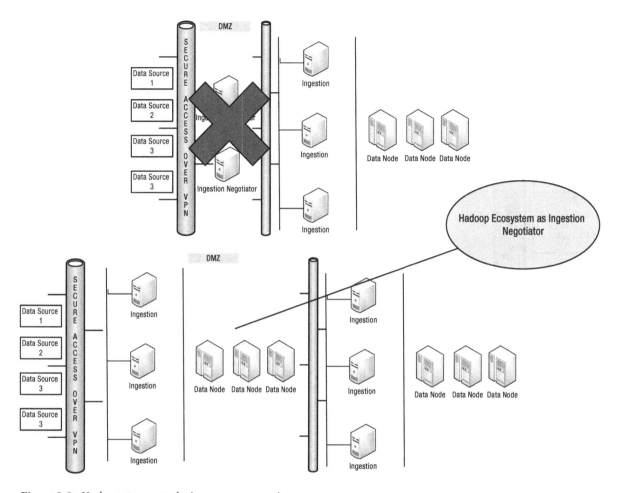

Figure 8-3. *Hadoop storage replacing resource negotiator*

Spine Fabric Pattern
Problem

How do you handle data coming in from multiple data sources with varying degrees of security implementation?

Solution

Implement a *spine fabric pattern* (Figure 8-4).

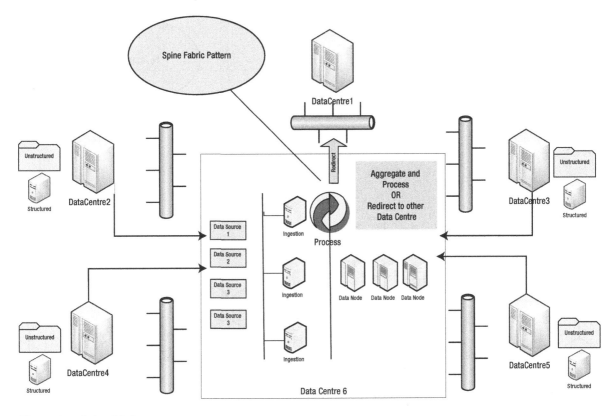

Figure 8-4. *Spine fabric pattern*

Data from multiple data centers, whether structured logs and reports or unstructured data, can be moved to a spine Hadoop ecosystem, which redirects the data to a target Hadoop ecosystem within the data center or in an external data center. The advantage of the spine fabric pattern is that the end data center is abstracted from the source data centers and new sources can be easily ingested without making any changes to the deployment pattern of the target or spine Hadoop ecosystems.

Federation Pattern
Problem

Can data from multiple sources be zoned and then processed?

Solution

Implement a *federation pattern* (Figure 8-5).

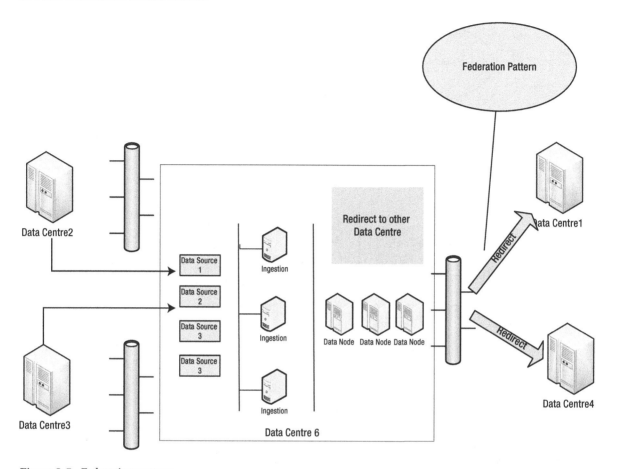

Figure 8-5. *Federation pattern*

In a federation pattern, the Hadoop ecosystem in the data center where the big data is processed splits and redirects the processed data to other Hadoop clusters within the same data center or in external data centers.

Lean DevOps Pattern
Problem

How can you automate infrastructure and cluster creation as virtual machine (VM) instances?

Solution

Implement a *Lean DevOps pattern* (Figure 8-6).

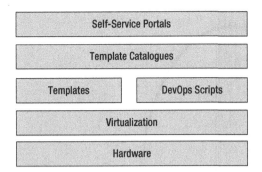

Figure 8-6. *Lean DevOps pattern*

Agile *Infrastructure-as-a-Code* (IaaC) scripts (exemplified by such products as Chef and Puppet) can be used to create templates of environment configurations and to re-create the whole virtual machine (VM) cluster and/or infrastructure as needed. Infrastructure teams and application teams might need to collaborate in the creation of instance and server templates to configure the applications and batch jobs properly.

Because every entity—whether it is a Hadoop component (such as a data node or hive metastore) or an application component (such as a visualization or analytics tool) has to be converted into a template and configured. Licenses for each component have to be manually configured. To be on the safe side, have the product vendor provide you with a virtual licensing policy that can facilitate your creation of templates. If the infrastructure is hardened in conformity with organizational policies, the vendor product might have to be reconfigured to ensure that it installs and runs successfully.

IBM's SmartCloud Orchestrator and OpsCode Chef are examples of popular DevOps implementation products. The Lean DevOps pattern has been successfully implemented by Netflix on Amazon Web Services Cloud.

Big Data on the Cloud and Hybrid Architecture
Problem

Can you avoid operational overheads by utilizing services available on the cloud?

Solution

Data residing in a data center can utilize the processing power of the cloud (Figure 8-7).

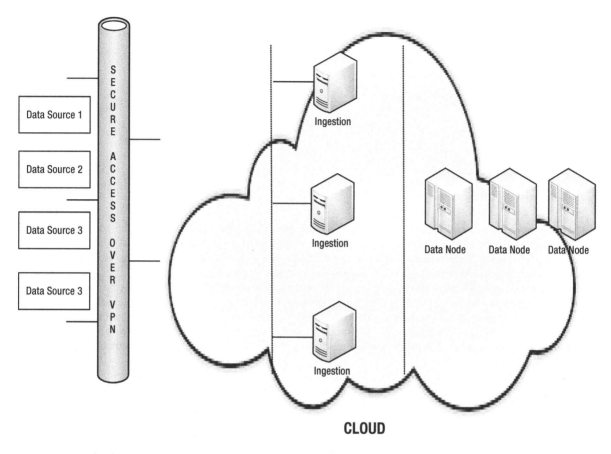

CLOUD

Figure 8-7. *Hybrid architecture*

There are several cloud options available for big data. AWS provides infrastructure as well as Amazon Elastic MapReduce (EMR). SAP provides the packaged software solution SAP OnDemand. IBM and Oracle provide the DevOps-based cloud implementations PureFlex/Maximo and Oracle Exalytics, which are offered in the form of self-service portals affording access to whole clusters along with the necessary software libraries. Such systems combine servers, storage, networking, virtualization, and management into a single infrastructure system.

Big Data Operations
Problem

Hadoop distributions provide dashboards to monitor the health of various nodes. Should a data center have an additional monitoring mechanism?

Solution

Hadoop distributions such as Cloudera Manager have been found to suffer occasional limitations in fully representing the health of various nodes because of failure by its agents. It is therefore advisable to supplement your Hadoop monitoring tools with end-to-end IT operations tools, such as Nagios or ZenOS.

Summary

Big data infrastructure presents a unique set of issues that can be addressed using the set of patterns discussed in this chapter. As big data infrastructure becomes more secure and newer packaged big data appliances and products emerge, new infrastructure and deployment patterns are continually emerging. Cloud-based solutions are becoming increasingly common, and clients increasingly elect hybrid architectures that cater to their pay-as-you-go needs.

CHAPTER 9

Big Data NFRs

Non-Functional requirements (NFRs) like security, performance, scalability, and others are of prime concern in big data architectures. Apart from traditional data security and privacy concerns, virtualized environments add the challenges of a hybrid environment. Big data also introduces challenges of NoSQL databases, new distributed file systems, evolving ingestion mechanisms, and minimal authorization security provided by the Hadoop-like platforms. Let's look at the different scenarios where NFRs can be fine-tuned by extending the patterns discussed in earlier chapters as well as new patterns.

"ilities"
Problem

What are the common "ilities" that we should be cognizant of while architecting big data systems?

Solution

Due to the distributed ecosystem, there are multiple "ilities" that a big data architect should consider while designing a system:

- **Reliability:** Reliability is the ability of the system to be predictable and give the desired results on time and every time. It is related to the integrity of the system to give consistent results to every user of the system.

- **Scalability:** The ability of the system to increase processing power within the same machine instance (vertical scaling) or to add more machine instances in parallel (horizontal scaling) is called *scalability*.

- **Operability:** Once in production, how amenable the system is for monitoring every aspect that could affect the smooth operations determines its operability.

- **Maintainability:** A system is highly maintainable if defects, change requests, and extended features can be quickly incorporated into the running system without affecting existing functionality.

- **Availability:** Metrics like 99.999 and 24*7 are used to define the availability of the system for the users so that there is no downtime or very little downtime.

- **Security:** *Distributed nodes, shared data, access ownership, internode communication*, and *client communication* are all prime candidates for security vulnerabilities that can be exploited in a big data system.

A big data architect has to provide support for all the aforementioned "ilities" and trade-off some of them against each other based on the application priorities.

Security

Traditional RDBMS systems have evolved over the years to incorporate extensive security controls like secure user and configuration management, distributed authentication and access controls, data classification, data encryption, distributed logging, audit reports, and others.

On the other hand, big data and Hadoop-based systems are still undergoing modifications to remove security vulnerabilities and risks. As during its inception, the primary job of Hadoop was to manage large amounts data; confidentiality and authentication were ignored.. Because security was not thought about in the beginning as part of the Hadoop stack, additional security products now are being offered by big data vendors. Because of these security concerns, in 2009 *Kerberos* was proposed as the authentication mechanism for Hadoop.

Cloudera Sentry, DataStax Enterprise, DataGuise for Hadoop, provide secure versions of Hadoop distributions. Apache Accumulo is another project that allows for additional security when using Hadoop.

The latest Hadoop versions have the following support for security features:

- Authentication for HTTP web clients

- Authentication with Kerberos RPC (SASL/GSSAPI) on RPC connections

- Access control lists for HDFS file permissions

- Delegation tokens for subsequent authentication checks after the initial authentication on Kerberos

- Job tokens for task authorization

- Network encryption

You have seen how to use ingestion patterns, data access patterns, and storage patterns to solve commonly encountered use-cases in big data architectures. In this chapter, you will see how some of these patterns can be further optimized to provide better capabilities with regard to performance, scalability, latency, security, and other factors.

Parallel Exhaust Pattern
Problem

How do I increase the rate of ingestion into disparate destination systems inside the enterprise?

Solution

The data ingested from outside the enterprise can be stored in many destinations, like *data warehouses, RDBS systems, NoSQL databases, content management systems,*and *file systems*. However, the speed of the incoming data passing through a single funnel or router could cause congestion and data regret. The data integrity also gets compromised due to leakage of data, which is very common in huge volumes of data. This data also hogs network bandwidth that can impede all other business-as-usual (BAU) transactions in the enterprise.

To overcome these challenges, an organization can adopt the *Parallel Exhaust pattern* (Figure 9-1). Each destination system has a separate router to start ingesting data into the multiple data stores. Each router, instead of publishing data to all sources, has a one-to-one communication with the destinations, unlike the "multi-destination" ingestion pattern seen in an earlier chapter. The routers can be scaled horizontally by adding more virtual instances in a cloud environment, depending on the volume of data and number of destinations.

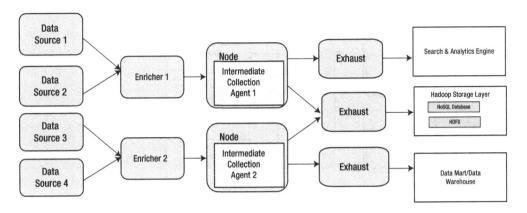

Figure 9-1. *Parallel Exhaust pattern*

Variety Abstraction Pattern
Problem

What is polyglot persistence?

Solution

One size does not fit all. With unstructured and structured data, for a business case, storage mechanisms can be a combination of storage mechanisms, like an RDBMS, a NoSQL database, and other forms of storage. For example, a web site can have transactional data in an RDBMS, session data stored in a key-value NoSQL database, relationships between users or products stored in a graph database, and so forth. Thus, the landscape database in an enterprise is moving toward a heterogeneous combination of different types of databases for different purposes.

Problem

With polyglot persistence (multiple data storage systems) becoming the norm in an enterprise, how do we make sure that we do not become tightly coupled to a specific big data framework or platform? Also, if we have to change from one product to another, how do we make sure that there is maximum interoperability?

Solution

With multiple analytics platforms storing disparate data, we need to build an abstraction of *application program interfaces* (APIs) so that data can be interchangeably transferred across different storage systems.

This helps in retaining legacy DW frameworks. This pattern (Figure 9-2) helps simplify the variety problem of big data. Data can be stored in, imported to, or exported to HDFS, in the form of storage in an RDBMS or in the form of appliances like IBM Netezza/EMC Greenplum, NoSQL databases like Cassandra/HO Vertica/Oracle Exadata, or simply in an in-memory cache.

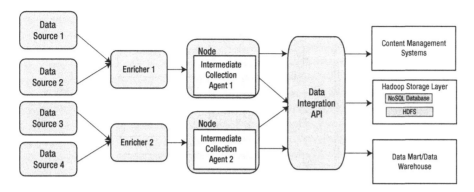

Figure 9-2. *Variety Abstraction pattern*

Real-Time Streaming Using the Appliance Pattern

Hadoop and MapReduce were created with off-line batch jobs in mind. But with the demand for real-time or near-real-time reports, abstraction of the data in a layer above the Hadoop layer that is also highly responsive requires a real-time streaming capability, which is being addressed by some big data appliances.

The other use-cases that require real-time monitoring of data are smartgrids, real-time monitoring of application and network logs, and the real-time monitoring of climatic changes in a war or natural-calamity zone.

Problem

I want a single-vendor strategy to implement my big data strategy. Which pattern do I go for?

Solution

The virtualization of data from HDFS to NoSQL databases is implemented very widely and, at times, is integrated with a big data appliance to accelerate data access or transfer to other systems. Also, for real-time streaming analysis of big data, appliances are a must (Figure 9-3).

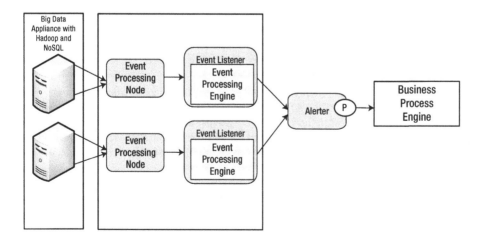

Figure 9-3. *Real-time streaming using an Appliance pattern*

EMC Greenplum, IBM PureData (Big Insights + Netezza), HP Vertica, and *Oracle Exadata* are some of the appliances that bring significant performance benefits. Though the data is stored in HDFS, some of these appliances abstract data in NoSQL databases. Some vendors have their own implementation of a file system (GreenPlum's OneFS) to improve data access.

Real-time big data analytics products like SAP HANA come integrated with appliances that are fine-tuned for maximum performance.

Distributed Search Optimization Access Pattern
Problem

How can I search rapidly across the different nodes in a Hadoop stack?

Solution

With the data being distributed and a mix of structured as well as unstructured, searching for patterns and insights is very time consuming. To expedite these searches, string search engines like Solr are preferred because they can do quick scans in data sources like log files, social blog streams, and so forth.

However, these search engines also need a "near shore" cache of indexes and metadata to locate the required strings rapidly. Figure 9-4 shows the pattern.

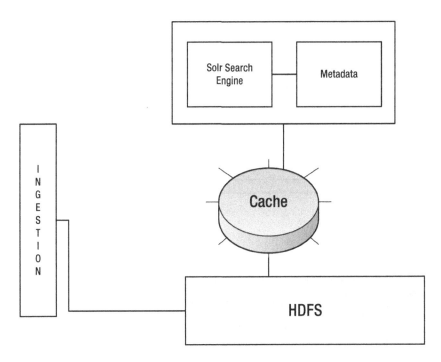

Figure 9-4. *Distributed search optimization Access pattern*

Anything as an API Pattern
Problem

If there are multiple data storages—that is, "polyglot persistence"—in the enterprise, how do I select a specific storage type?

Solution

For a storage landscape with different storage types, data analysts need the flexibility to manipulate, filter, select, and co-relate different data formats. Different data adapters should also be available at the click of a button through a common catalog of services. The Service Locator pattern, where data-storage access is available in a SaaS model, resolves this problem. However, this pattern, which simplifies interoperability and scalability concerns for the user, is possible only if the underlying platform is API enabled and abstracts all the technical complexities from the service consumer (Figure 9-5).

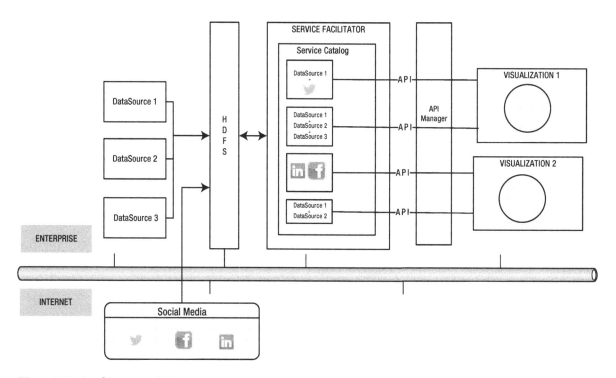

Figure 9-5. *Anything as an API pattern*

Security Challenges
Problem

What are the key security challenges in big data clusters?

Solution

Here are the key security challenges you'll face with big data clusters:

- The distributed nature of the data and the autonomy that each node needs to have to process, manage, and compute data creates a multi-armed Hydra that is difficult to secure against unauthorized or unauthenticated access.

- Since the big data cluster is a redundant architecture, with data split and replicated across multiple servers, after a period of time managing the multiple copies of data becomes a very complex governance problem.

- Since each node is independent and a peer to others, the access ownership is mostly at schema level. Also, since NoSQL databases do not have referential integrity and validations at the database level, the user applications have to build the validations in the UI or business layer.

- Due to the use of multiple nodes, there are frequent handoffs between nodes. These handoff points become prime candidates for man-in-the-middle attacks because most communication between nodes is via RPC.

- Since all nodes are peers, it is difficult to define a "gateway" where a DMZ or firewall can be set up.

- Node coordinating and access control are managed by frameworks like *Zookeeper* or *YARN*, whose main concern is to detect failure of nodes and switch to live nodes. They are not designed with enterprise security in mind.

Operability
Problem

What are the key operational challenges for big data administrators?

Solution

Following are the most important operational challenges for big data administrators:

- Big data administrators need the right tools to manage the mammoth clusters.

- Replication should be authorized properly using Kerberos; otherwise, any rogue client can create their own copy of the data on any of the nodes.

- Data has to be encrypted, but at the same time, it should not lead to a lag in decryption, as a very small lag can multiply exponentially across multiple nodes.

- Since the big data environment is essentially a megacluster that could be spread across multiple physical locations, it needs multiple administrators with different access rights and a clear separation of responsibilities.

- Since it is possible for different nodes to have different releases of OS, VM, and so forth, a governance tool is needed that manages these configurations and patches.

- We know Kerberos is used to authenticate Hadoop users. However, if a Kerberos ticket is stolen by a man-in-the-middle attack, a client can be duplicated and a rogue clone can be added to the cluster.

- Distributed logging is a must to track the train errors across the cluster landscape. Open source tools like scribe and logstash provide good log management in big data environments.

- As more and more big data services become API driven, strong API management tools are required that can keep track of REST-based APIs and provide life-cycle management for these APIs.

Big Data System Security Audit
Problem

As a big data compliance and security auditor, what are the basic questions that I should ask the concerned stakeholders of the company?

Solution

The following set of questions is a good starting point to begin a big data related audit:

- What are the various technology frameworks being used in the big data ecosystem for computation, data access, pattern recognition, task, and job management and monitoring?

- Who are the primary user-groups running big data queries? What functions are they trying to perform using these queries?

- Are these big data queries made by authorized users in real time or in batch mode using map reduce jobs?

- Are there any new applications built specifically to leverage the big data functions? Have those applications been audited?

- What is the amount of replication of data across nodes, and is there a complete dependency chart that is updated in real time for all these nodes?

- What tools and processes are being used for statistical analysis, text search, data serialization, process coordination, and workflow and job orchestration?

- What is the process for replication and recovery across node clusters?

- What are the different distributed file systems in the big data environment? Are they all storing structured data or unstructured?

- What are the backup, archiving, and recovery processes for all the above applications and storage systems?

- Are there any caching systems both persistent and in-memory? What is the level of confidentiality of the data stored in the caches? Are they purged regularly, and is there protection from other unauthorized access during run-time?

- Are there any performance criteria established as baselines, and what is the process to detect noncompliance and solve noncompliance issues? Typical reasons for noncompliance could be large input records, resource contention (CPU, network, and storage), race conditions between competing jobs, and so forth.

- How are different versions of Hadoop, OS, VMs, and so forth tracked and managed?

Big Data Security Products

Problem

If I want to harden my big data architecture, are there any open source, common off the shelf (COTS) products I can buy?

Solution

Cloudera's Sentry is an open source, enterprise-grade, big data security and access-control software that provides authorization for data in Apache Hadoop. It can integrate with Apache Hive.

Problem

Is there a big data API management tool I can use with minimum loss in performance?

Solution

Intel Expressway API Manager (Intel EAM) is a security-gateway enforcement point for all REST Hadoop APIs. Using this manager, all Hadoop API callers access data and services through this gateway.

It supports the authentication of REST calls, manages message-level security and tokenization, and protects against denial of service. There are other products also in the market that can do this.

Problem

I want to maintain the confidentiality of my data and only expose data relevant to a user's access levels. What should I use?

Solution

InfoSphere Optim data masking (InfoSphere Optim DM) is a product that ensures data privacy, enables compliance, and helps manage risk. Flexible masking services allow you to create customized masking routines for specific data types or leverage out-of-the-box support.

InfoSphere Optim data masking on demand is one of the many masking services available for Hadoop-based systems. You can decide when and where to mask based on your application needs.

Problem

For efficient operability, I need good monitoring tools for big data systems. Are there any COTS products I can buy?

Solution

Products like Infosphere *Guardium* monitor and audit high-performance, big data analytics systems. InfoSphere Guardium provides built-in audit reporting to help you demonstrate compliance to auditors quickly.

IBM Tivoli Key Lifecycle Manager (TKLM) is another product that enhances data security and compliance management with a simple and robust solution for key storage, key serving, and key lifecycle management for IBM self-encrypting storage devices and non-IBM devices. TKLM offers an additional layer of data protection by providing encryption lifecycle key management for self-encrypting storage devices above and beyond the Guardium and Optim security capabilities.

Problem

What are the future plans of the Hadoop movement to enhance data protection of the Hadoop ecosystem?

Solution

Project Rhino is an open source Hadoop project that's trying to address security and compliance challenges. Project Rhino is targeted to achieve the following objectives:

- Framework support for encryption and key management

- A common authorization framework for the Hadoop ecosystem

- Token-based authentication and single sign-on

- Extend HBase support for ACLs to the cell level

- Improve audit logging

Problem

What are some of the well-known, global-regulatory compliance rules that big data environments spread over different geographies and over public and private clouds have to comply with?

Solution

Table 9-1 shows some of the significant compliance rules that affect big data environments, showing what the constraint is along with the jurisdiction.

Table 9-1. *Big Data Compliance Issues*

Compliance Rule	Data Constraints	Geography
EU Model Clauses for data transfers outside the EU	Allows clients with EU data to lawfully use U.S. data centers.	This is a basic requirement for any client with consumer or employee data of EU origin.
A HIPAA "**Business Associate Agreement**"	Allows clients with HIPAA-regulated data to lawfully use the data center.	U.S. insurers, healthcare providers, and employee health benefit plans (often relevant for HR work)
Model **Gramm-Leach-Bliley** insurance regulations	Enables U.S. insurers and other financial institutions[1] to use the data center to host consumer data.	U.S. insurers and other financial institutions, relative to consumer data.[2]
Massachusetts Data Security Regulations	Allows personal data from Massachusetts to be hosted by the data center consistent with MA requirements, which are currently the strictest state-level standards in the U.S.	Companies with personal data of Massachusetts origin.
The UK Data Protection Act	Allows data from UK to be hosted by the data center consistent with local UK requirements.	Companies with personal data of UK origin.

(continued)

Table 9-1. (*continued*)

Compliance Rule	Data Constraints	Geography
Spanish Royal Decree on information security	Allows data from Spain to be hosted by the data center consistent with local Spanish requirements.	Companies with personal data of Spanish origin.
Canadian federal PIPEDA **and provincial** PIPA acts.	Allows clients with personal data of Canadian origin to use the data center.	Companies with personal data of Canadian origin.
Canadian provincial statutes focused on health data	Allows for data center hosting of health-related data from relevant Canadian provinces.	Entities with health-related data of Canadian origin.

1

2

Summary

Big data architectures have to consider many non-functional requirements at each layer of the ecosystem. There are a host of tools to support administrators, developers, and designers in meeting these service level agreements (SLAs). Various design patterns can be used to improve the "ilities" without affecting the functionality of the use-case. The horizontal cross-cutting concerns can be addressed by the appropriate mix of design patterns, tools, and processes.

CHAPTER 10

Big Data Case Studies

This chapter examines how the various patterns discussed in previous chapters can be applied to business problems in different industries. To arrive at the solution of a given business problem, architects apply combinations of patterns across different layers of the entire application architecture as appropriate to the unique business requirements and priorities of the problem at hand. The following case studies exemplify how architects combine patterns to solve particular business problems.

Case Study: Mainframe to Hadoop-Based NoSQL Database

Problem

A financial organization's current data warehouse solution is based on a legacy mainframe platform. This solution is becoming very expensive as more and more data gets generated every day. Moreover, because the databases supported are legacy formats (such as line IMS and IDMS), it is not easy to transform and merge this data with the other data sources in the enterprise for joint analytical processing. The CIO is looking for a less expensive and more current platform.

Solution

The CIO concluded that migrating the legacy data to a NoSQL-based platform (such as HP Vertica) would provide the following benefits:

- A higher level of data compression, providing lower storage costs and improved performance

- A native data load option, avoiding the need to use a third-party ELT tool

- Easier integration

- Better co-analysis of data from multiple data sources in the organization

Figure 10-1 shows the patterns implemented in migrating to a NoSQL platform.

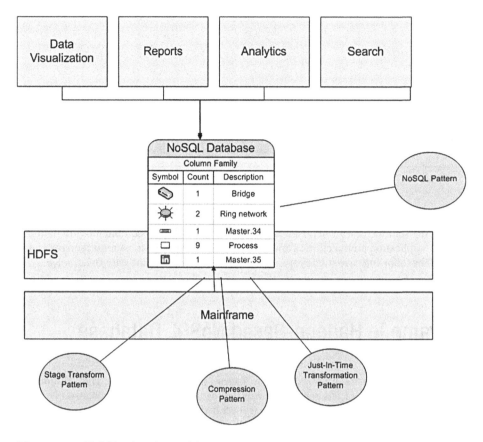

Figure 10-1. *NoSQL migration architecture*

Examples of technologies used include the following:

- HP Vertica

- VSQL (for Native ELT: Extract, Load, Transform)

- AutoSys (for scheduling)

- Unix Shell/Perl scripting

Table 10-1. *Patterns implemented in the Mainframe to Hadoop case study*

Pattern Type	Pattern Name
Big data storage pattern	NoSQL Pattern
Ingestion and streaming pattern	Just-In-Time Transformation Pattern
Analysis and visualization pattern	Compression Pattern
Big data access pattern	Stage Transform Pattern

Case Study: Geo-Redundancy and Near-Real-Time Data Ingestion

Problem

A high-tech organization has multiple applications spread geographically across multiple data centers. All application usage logs have to be synchronized with every data center for near-real-time analysis. The current implementation of the RDBMS is capable of providing replication across data centers, but it is very expensive and the cost is increasing as more data accumulates every day. What cost-efficient solution would enable active-active geo-redundant ingestion across data centers to address failover and provide more near-real-time access to data?

Solution

The big data architects choose an open-source (hence low-cost) NoSQL-based platform (such as Cassandra) that can be configured for fast data synchronization and replication across data centers, high availability, and a high level of data compression for lower storage costs and improved performance. This solution provides very high, terabyte-scale ingestion rates across data centers.

Figure 10-2 shows the patterns implemented in changing to a geo-redundant NoSQL-based platform.

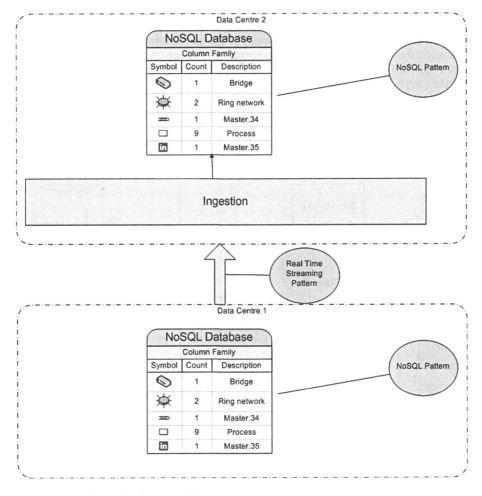

Figure 10-2. *Geo-redundancy architecture*

Table 10-2. *Patterns implemented in the Geo-Redundancy case study*

Pattern Type	Pattern Name
Big data storage	NoSQL Pattern
Ingestion and streaming pattern	Real-Time Streaming Pattern

Case Study: Recommendation Engine
Problem

An organization has an existing recommendation engine, but it is looking for a high-performing recommendation engine and reporting tool that can handle its increasing volumes of data. The existing implementation is based on a subset of the total data and hence is failing to generate optimal recommendations. What high-performing recommendation engine could look at the current volume data in its totality and scale up to accommodate load increases going forward?

Solution

The organization's combinatory solution is to move to a Hadoop-based storage mechanism (providing increased capacity), a NoSQL-based Cassandra database for real-time log-processing (providing higher-speed data access), and an R-based solution for machine-oriented learning.

Figure 10-3 shows the patterns implemented to enable real-time streaming for machine learning.

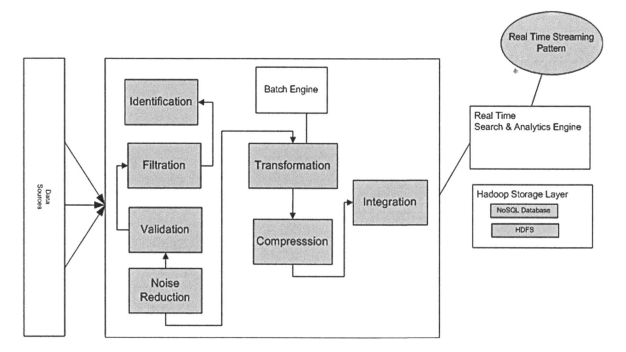

Figure 10-3. *Real-time streaming for machine-learning architecture*

Table 10-3. *Pattern implemented for the Recommendation Engine case study*

Pattern Type	Pattern Name
Ingestion and streaming pattern	Real-Time Streaming Pattern

Examples of technologies used include the following:

- Cassandra

- HDFS, Hive, HBase, Pig, Hive

- Map-R

Case Study: Video-Streaming Analytics

Problem

A telecommunication organization needs a solution for analyzing customer behavior and viewing patterns in advance of a rollout of video-over-IP (VOIP) offerings. The logs have to be compared to region-specific, feature-specific existing system data spread across multiple applications. Because the volume of data is already huge and the VOIP logs data will add many terabytes, the organization is looking for a robust solution to apply across all devices and systems.

Solution

The CTO chooses a Hadoop-based big data implementation capable of storing and analyzing the huge volume of raw system data and scaling up to accommodate the VOIP metadata: namely, a consolidated log-access, log-parse, and analysis platform that is able to transform data using Pig, to store data in HDFS and NoSQL MongoDB, and to incorporate machine-learning tools for analytics.

Figure 10-4 shows the patterns implemented to enable video-streaming analytics.

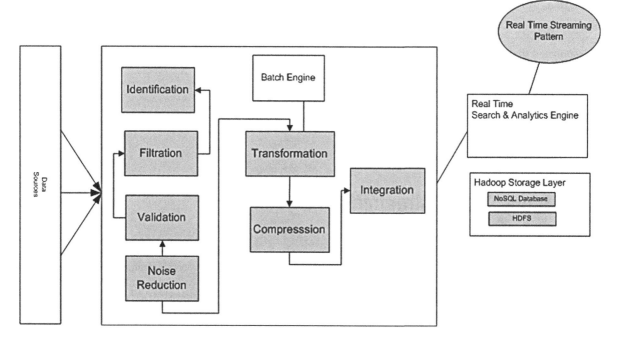

Figure 10-4. *Video analytics architecture*

Table 10-4. *Pattern implemented for the Video Analytics case study*

Pattern Type	Pattern Name
Ingestion and streaming pattern	Real-Time Streaming Pattern

Examples of technologies used include the following:

- Hadoop
- Python
- Memcache
- Jetty, Apache
- Web/Mobile Dashboards/Analytics
- Amazon EMR

Case Study: Sentiment Analysis and Log Processing
Problem

An existing ecommerce organization experienced system failures and data inconsistencies during the holiday season. Major issues included penalties tied to performance-based service-level agreements (SLAs)s. The organization is looking for a new platform that could take the holiday season load, help them avoid penalties, and ensure customer satisfaction.

Solution

The company decided to set up a big data platform with Hadoop and Hive to enable web and application server historic and real-time log analysis: namely, a NoSQL-based solution (such as MongoDB) for analyzing the application logs and an R-based machine-learning engine and visualization tool (such as Tableau) for better viewing of requests, faster resolution of defects, reduced down time, and better customer satisfaction.

Figure 10-5 shows the patterns implemented to enable scalable sentiment analysis and log processing.

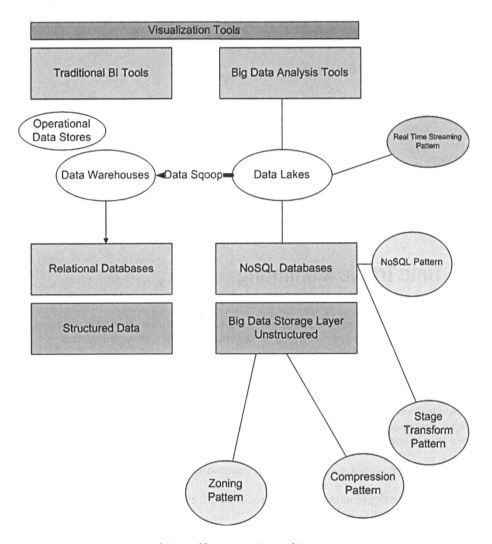

Figure 10-5. *Sentiment-analysis and log-processing architecture*

Table 10-5. *Patterns implemented for the Sentiment Analysis case study*

Pattern Type	Pattern Name
Ingestion and streaming pattern	Real-Time Streaming Pattern
Big data analysis and visualization pattern	Zoning Pattern Compression Pattern
Big data access pattern	Stage Transform Pattern
Big data storage	NoSQL Pattern

Examples of technologies used include the following:

- HDFS, Hive, HBase

- NoSQL - MongoDB.

- R

- Log Data Processing

- MapReduce

- Compuware DynaTrace

- Data Analytics – Tableau

Case Study: Real-Time Traffic Monitoring
Problem

An organization wants to create a real-time traffic analysis and prediction application that can be used to control traffic congestion and streamline traffic flow. The application must be targeted to provide cost optimization in commuting and help reduce waiting time and pollution levels.

Data has to be captured from existing government-provided datasets that include sources such as traffic-camera, traffic-sensor, GPS, and weather-prediction systems. The government data needs to be coupled with social media to assist in predicting traffic speed and volume on roads.

The analysis scenarios include the following:

Analysis of historical data to gain insights and understand patterns of behavior of traffic and road incidents

Prediction of traffic speed and volume well ahead of time, based on analysis of real-time and historical traffic data

Prediction of alternate cost-effective commute paths by analyzing situational traffic conditions across the entire transportation network

The application needs to provide a catalog of services based on social media, governmental data, and different dataset options.

Solution

The organization decided to set up a big data platform using Hadoop, an abstracted layer of data above HDFS in the form of HP Vertica, and a visualization tool. The organization opted to use the cloud-based Amazon Web Service for storage and analytics.

Multiple patterns are applied at various layers of the architecture, as depicted in Figure 10-6. The patterns shown in that figure were used to enable monitoring of traffic in real time.

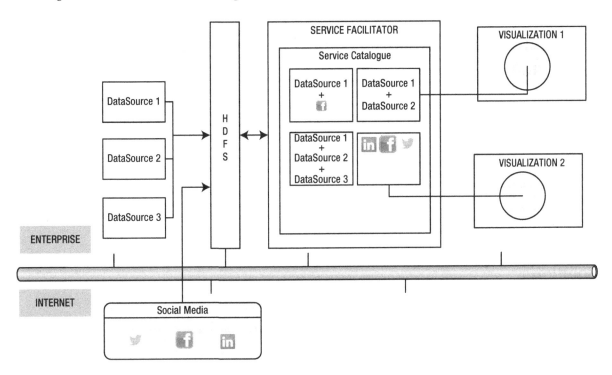

Figure 10-6. *Traffic-monitoring architecture*

Table 10-6. *Patterns implemented for Traffic Monitoring*

Pattern Type	Pattern Name
Ingestion and streaming pattern	Real-Time Streaming Pattern
Big data analysis and visualization pattern	Zoning Pattern Compression Pattern
Big data access pattern	Service Locator Pattern
Big data storage pattern	NoSQL Pattern
NFR patterns	Distributed Search Optimization Access Pattern

Examples of technologies used include the following:

- Hadoop

- HP Vertica

- Web/Mobile Dashboards/Analytics

- Amazon Web Services

Case Study: Data Exploration for Suspicious Behavior on a Stock Exchange

Problem

A financial organization processes millions of order entries per day. Whenever online statistical surveillance models identify suspicious behavior, the organization wants to have enhanced capability to gather data pertinent to the suspicious behavior as quickly and cheaply as possible.

The solution needs to be able to do the following:

- Integrate social media data with historical orders and trades

- Gather information from other sources within the organization

- Present this information in an integrated fashion

Solution

The lead architect applied the patterns mentioned in Figure 10-7. The solution is based on Hadoop, Storm, Flume, and IBM Netezza. DataStax Cassandra acted as the NoSQL database to enable real-time analysis.

Figure 10-7 shows the patterns implemented to enable data forensics on a stock exchange.

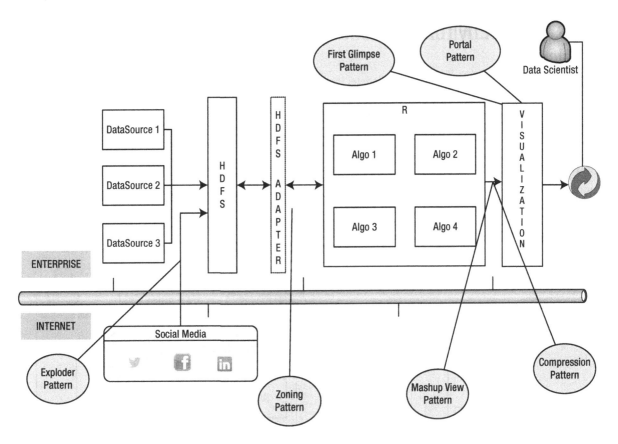

Figure 10-7. *Data forensics on a stock exchange*

Table 10-7. *Patterns implemented for the Data Forensics case study*

Pattern Type	Pattern Name
Ingestion and streaming pattern	Real-Time Streaming Pattern
Big data analysis and visualization pattern	Zoning Pattern Compression Pattern
Big data access pattern	Service Locator Pattern
Big data storage	NoSQL Pattern

Examples of technologies used include the following:

- Hadoop

- IBM Netezza

- DataStaX Cassandra

- Tableau

- R

Case Study: Environment Change Detection

Problem

An institute wants to build an application that detects environmental changes to water resources in real time. The application has to source data from multiple data sources (such as sensor and meteorological sources) hosted in various environmental institutes and government departments. The data has to be presented to scientists and energy analysts for real-time monitoring of the water resources and environmental data.

Solution

The CTO chooses an all-IBM big data platform with IBM BigInsights, IBM InfoSphere Streams, and IBM Vivisimo as the technologies applied against the patterns shown next.

Figure 10-8 shows the patterns implemented to enable environment change detection.

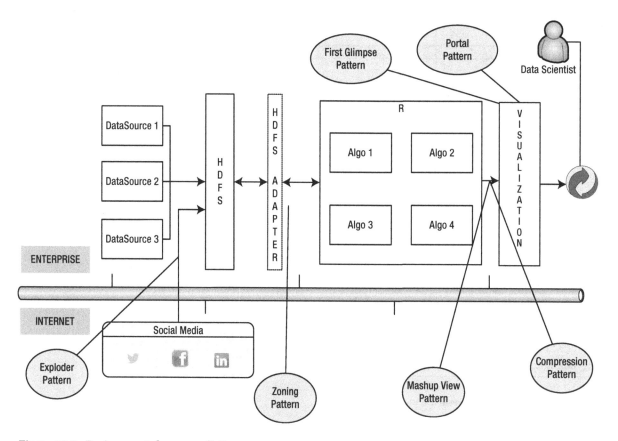

Figure 10-8. *Environment change prediction*

Table 10-8. *Patterns implemented in Environment Change Prediction*

Pattern Type	Pattern Name
Ingestion and streaming pattern	Real-Time Streaming Pattern
Ingestion and streaming pattern	Just-In-Time Transformation Pattern
Analysis and visualization patterns	Compression Pattern
Big data access pattern	Stage Transform Pattern

Examples of technologies used include the following:

- IBM Vivisimo
- IBM BigInsights
- IBM Cognos

Summary

A multitude of practical business, academic, financial, and scientific problems are susceptible to solution using big data architectures. The patterns described in this book can be applied to all the layers of your big data architecture. The rapid pace of technological advances in tools and products ensures the continual emergence of new patterns, new variants of existing patterns, and new combinations of patterns in increasingly industrialized out-of-the box solutions.

CHAPTER 11

Resources, References, and Tools

There is a plethora of big data products available from large and small vendors. Some of these products cater to niche areas like social media analytics and NoSQL databases, while some have a *Hadoop* ecosystem with a combination of infrastructure, visualization, and analytical capabilities. This chapter gives a broad overview of many of the products you will need to implement the architecture and patterns described in this book.

Big Data Product Catalog

Problem

List the main big data product areas and the associated vendors of those products.

Solution

Table 11-1 lists tools you can use as a solution, as well as the vendors providing those tools.

Table 11-1. *Distributed and Clustered Flume Taxonomy*

Tools	Vendors
Hadoop Distributions	Cloudera
	Hortonworks
	MapR
	IBM BigInsights
	Pivotal HD
	Microsoft Windows Azure cloud platform, HDInsight
In-Memory Hadoop	Intel
Hadoop Alternatives	HPCC Systems from LexisNexis
Hadoop SQL Interfaces	Apache Hive
	Cloudera Impala
	EMC HAWQ
	Hortonworks Stinger

(*continued*)

Table 11-1. (*continued*)

Tools	Vendors
Ingestion Tools	Flume
	Storm
	S4
	Sqoop
Map Reduce Alternatives	Spark
	Nokia Disco
Cloud Options	AWS EMR
	Microsoft Azure
	Google BigQuery
NoSQL Databases	IBM Netezza
	HP Vertica
	Aster Teradata
	Google BigTable
	EMC Greenplum
In-Memory Database Management Systems	SAP HANA
	Oracle Exalytics
Visualization	Tableau
	QLikView
	Tibco Spotfire
	MicroStrategy
	SAS VA
Search	Solr
Analytics	SAS
	Revolution Analytics
	Pega
Integration Tools	Talend
	Informatica
Operational Intelligence Tools	Splunk
Graph Databases	Neo4J
	OpenLink
Document Store Database Management Systems	MongoDB
	Cloudant
	MarkLogic
	Couchbase
Datasets	InfoChimps
Social Media Integrator	Clarabridge
	radian6
	SAS
	Informatica PowerExchange
Archive Infrastructure	EMC Ipsilon

(*continued*)

Table 11-1. (*continued*)

Tools	Vendors
Data Discovery	IBM Vivisimo
	Oracle Endeca
	MarkLogic
Table-Style Database Management Services	Cassandra
	HP Vertica
	DataStax
	Teradata

Hadoop Distributions
Problem

Apache Hadoop does not come integrated with all the components required for an enterprise-scale big data system. Do I have any better options to save the time and effort to configure multiple frameworks?

Solution

Because the Hadoop ecosystem is made up of multiple entities (Hive, Pig, HDFS, Ambari, and others), with each entity maturing individually and coming up with newer versions, there are chances of version incompatibility, security, and performance-tuning issues. Vendors like Cloudera, MapR and Hortonworks do a good job to package it all together into one distribution and manage the incompatibility, performance, and security issues within the packaged distribution. This greatly helps because the maintenance support for these open source entities is only through forums.

These vendors are coming up with their own SQL interfaces and monitoring dashboards and contributing back to the Apache Hadoop community, thereby enriching the open source community. Here are some examples:

- MapR stands out as a unique file system. Unlike the HDFS file systems in Apache Hadoop, MapR allows you to mount the cluster as an NFS volume. "HDFS" is replaced by "Direct Access NFS." Further details can be found at http://www.mapr.com/Download-document/4-MapR-M3-Datasheet.

- Hortonworks stands out for its offering on Windows operating systems. Hortonworks is the only vendor providing Hadoop on Windows OS.

- Intel provides Hadoop storage that is *in-memory*.

A typical packaged distribution covers all of the open source Hadoop entities.

In-memory Hadoop

Intel provides optimization for solid state disks and cache acceleration. See www.intel.com/bigdata for information on Intel's big data resources in general.

Over and above the core open source Hadoop entities, vendors provide additional services such as those shown in Table 11-2.

Table 11-2. *Vendor-Specific Hadoop Services*

	Cloudera	Hortonworks	MapR	EMC
Operations	Cloudera Manager			
Security	-	Knox Gateway		
Massively Parallel Processing (MPP) query engine	Impala	Stinger	Drill	HAWQ

Hadoop Alternatives
Problem

Is Apache Hadoop the only option to implement big data, map reduce, and a distributed file system?

Solution

The nearest open source alternative to Hadoop is the HPCC system. Refer to `http://hpccsystems.com/Why-HPCC/How-it-works`.

Unlike Hadoop, HPCC provides massively parallel processing and a shared nothing architecture not **based on any type of key-value** NoSQL databases. See `http://hpccsystems.com/Why-HPCC/case-studies/lexisnexis` for a case study of HPCC implementation in LexisNexis document management.

Hadoop SQL Interfaces
Problem

How can I improve the performance of my Hive queries?

Solution

Apache Hive is the most widely used open source SQL interface. Apache Hive can run over HDFS or over the NoSQL HBase columnar database. Apache Hive was developed to make a SQL developer's life easier. Instead of forcing developers to learn a new language or learning new CLI commands to run MapReduce code, Apache Hive provides a SQL-like language to trigger map reduce jobs.

Cloudera Impala, EMC HAWQ, and Hortonworks Stinger are some of the products available that can overcome the performance issues encountered while using Apache Hive. Hortonworks Stinger is relatively new to the market. New aggregate functions, optimized query, and optimized Hive runtime are some of the features added in these products.

There are many resources you can consult to find comparisons of EMC HAWQ with Cloudera Impala and other such products. One that you may find useful is at: `https://www.informationweek.com/software/information-management/cloudera-impala-brings-sql-querying-to-h/240153861`. You find more information going to `http://www.cloudera.com/content/cloudera/en/products/cdh/impala.html` and `http://hortonworks.com/blog/100x-faster-hive`.

Ingestion tools

Problem

What are the essential tools/frameworks required in your big data ingestion layer?

Solution

There are many product options to facilitate batch-processing-based ingestion. Here are some major frameworks available:

- **Apache Sqoop:** A tool used for transferring bulk data from RDBMS to Apache Hadoop and vice-versa. It offers two-way replication, with both snapshots and incremental updates.

- **Chukwa:** Chukwa is a Hadoop subproject that is designed for efficient log processing. It provides a scalable distributed system for monitoring and analyzing log-based data. It supports appending to existing files and can be configured to monitor and process logs that are generated incrementally across many machines.

- **Apache Kafka:** A distributed publish-subscribe messaging system. It is designed to provide high-throughput persistent messaging that's scalable and allows for parallel data loads into Hadoop. Its features include the use of compression to optimize I/O performance and mirroring to improve availability and scalability and to optimize performance in multiple-cluster scenarios. It can be used as the framework between the router and Hadoop in the multi-destination pattern implementation.

- **Flume:** A distributed system for collecting log data from many sources, aggregating it, and writing it to HDFS. It is based on streaming data flows. Flume provides extensibility for online analytic applications. However, Flume requires a fair amount of configuration, which can become complex for very large systems.

- **Storm:** Supports event-stream processing and can respond to individual events within a reasonable time frame. Storm is a general-purpose, event-processing system that uses a cluster of services for scalability and reliability. In Storm terminology, you create a topology that runs continuously over a stream of incoming data. The data sources for the topology are called *spouts*, and each processing node is called a *bolt*. Bolts can perform sophisticated computations on the data, including output to data stores and other services. It is common for organizations to run a combination of Hadoop and Storm services to gain the best features of both platforms.

- **InfoSphere Streams:** Performs complex analytics of heterogeneous data types. InfoSphere Streams can support all data types. It can perform real-time and look-ahead analysis of regularly generated data, using digital filtering, pattern/correlation analysis, and decomposition, as well as geospacial analysis.

- **Apache S4:** A real-time data ingestion tool used for processing continuous streams of data. Client programs that send and receive events can be written in any programming language. S4 is designed as a highly distributed system. Throughput can be increased linearly by adding nodes into a cluster.

Map Reduce alternatives

Problem

For multinode parallel processing, is MapReduce the only algorithm option?

Solution

Spark and Nokia DISCO are some of the alternatives to MapReduce. A fair comparison can be found at
`http://www.bytemining.com/2011/08/Hadoop-fatigue-alternatives-to-Hadoop/`.

Because most vendor products and enhancements are focused on MapReduce jobs, it makes sense to stick to
MapReduce unless there is a pressing need to look for a massive parallel processing option.

Cloud Options

Problem

Buying inexpensive hardware for large big data implementations can still be a very large capital expense. Are
there any pay-as-you-go cloud options?

Solution

Amazon EMR is a public cloud based web service that provides huge data computing power. Amazon EMR uses
Amazon S3 for storage unlike the Hadoop HSFs storage. Amazon EMR data ingestion, analysis and import/export
concerns are different from a typical HDFS based Hadoop system.

Amazon EMR data ingestion, analysis, and import/export concerns are discussed at
`http://media.amazonwebservices.com/AWS_Amazon_EMR_Best_Practices.pdf`.

Table-Style Database Management Services

Problem

My existing database and analytics experts have RDBMS and SQL skills. How can I quickly make them big data
users?

Solution

Cassandra, HP Vertica, DataStax, Oracle Exadata, and Aster Teradata are some table-style, database-management
services that run over Hadoop and provide the abstraction needed to reduce latency and improve performance.

Compared to NoSQL DBMSs, table-style DBMSs bring vendor lock-in, hardware dependencies, and complicated
clustering. Table-style DBMSs also hog memory and network, and hence have consistency issues across clusters. Data
integrity and consistency suffer as performance improves.

Some table-style DBMSs come packaged as a combo of hardware, platform, and software. Hence, the eventual
cost and ROI need to be thoroughly investigated before opting for it.

You can find information concerning the total cost of ownership (TCO) for moving to Oracle Exadata at the
following web site: `http://www.zdnet.com/reproducing-youtube-on-oracle-exadata-1339318266/`.

NoSQL Databases

Problem

Is there a "one solution fits all" NoSQL database?

Solution

NoSQL databases (Figure 11-1) are very use-case-centric, unlike RDBMS, which are generic and cater to multiple system needs. Maintenance of data integrity and consistency can become a concern in the long run.

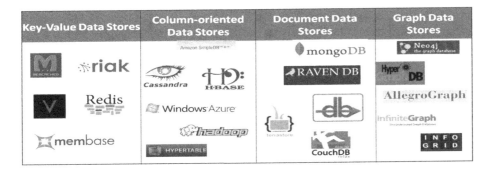

Figure 11-1. *NoSQL databases*

Also, there are multiple key-value pair, graph, and document databases that have very different implementations. Moving from one to another might raise issues.

In-Memory Big Data Management Systems

Problem

Real-time big data analysis requires in-memory processing capabilities. Which are the leading products?

Solution

SAP HANA and Oracle Exalytics are some of the leading products for in-memory processing. Oracle and SAP HANA provide their own set of adapters to connect to various other products, as well. Though different as far as hardware and platform, both products provide comparable features. The only concern is vendor lock-in and the inability to integrate with any other NoSQL database. Implementing these vendor products might require significant changes to a customer's existing high-availability and disaster-recovery processes.

DataSets

Problem

Are there any large data sets available in the public domain that I can use for my big data pilot projects?

Solution

Data.gov is an official US government web site. You can visit `http://catalog.data.gov/dataset` to see about 100,000 datasets belonging to different categories.

Data Discovery

Data discovery and search capabilities have gained more importance because of the new non-enterprise and social-media data that has started feeding in to an organization's decision-making systems. Discovering insights from unstructured data (videos, call center audios, blogs, twitter feeds, Facebook posts) has been a challenge for existing decision-making systems and has given birth to new products like IBM Vivisimo, Oracle Endeca, and others.

You can find more information at the following web sites: `https://wikis.oracle.com/display/endecainformationdiscovery/Home;jsessionid=7EF303D9FADB3215001F27A4F4DACFE5` and `http://www.ndm.net/datawarehouse/IBM/infosphere-dataexplorer-vivisimo`.

Visualization

Problem

There are so many cool visualization tools coming up every day, how do I select the appropriate tool for my enterprise?

Solution

High-volume, real-time analytics has brought in newer products in the market like the following ones:

- Tableau

- QLikView

- Tibco Spotfire

- MicroStrategy

- SAS VA

From a market presence perspective, QLikView has been around for a while. Comparatively, Tableau is a new entrant into the market. SAS has also jumped into the fray with its SAS Visual Analytics offering. MicroStrategy and Tibco Spotfire also have a substantial market presence.

Though these are in-memory visualization tools that are highly integrated with the Hadoop ecosystem, issues such as the following exist with them:

- Data is expected in a certain format for better performance.

- Integration with Apache Hive is at times not of very high performance.

- Apache Hive has to be modified to provide data in an aggregated manner to get high performance.

Analytics Tools

Here are some of the available analytics tools:

- Revolution R Analytics
- SAS

Though Revolution R Analytics is a new entrant into the market, it has made an immense impact. Because SAS is the old horse, it has an advantage over Revolution Analytics. However, it might be too costly an affair in the long run.

Data Integration Tools

Problem

I already invested in business intelligence (BI) tools like Talend, Informatica, and others. Can I use them for big data integration?

Solution

Talend, Informatica, Pentaho, and IBM DataStage are some of the tools that can act as ETL tools, as well as scheduling tools.

Some tools are still not mature and do not have all the adapters and accelerators needed to connect to all the Hadoop ecosystem entities. There might be restrictions on which versions of the entities these might work with.

Not all the tools are compatible with different distributions of Hadoop (like Cloudera, Hortonworks, MapR, Intel, and IBM BigInsights).

You can read about the Talend features at `http://www.talend.com/products/big-data/matrix`.

You can find additional information about Informatica PowerExchange at `http://www.informatica.com/us/products/enterprise-data-integration/powerexchange/`.

Summary

Since the big data industry is still evolving, there might be more products that will emerge as leaders in the field. As I am writing this, Oracle has launched Big Data X4-2 appliance, Pivotal has launched a Platform as a Service (PaaS) called "Pivotal One," and AWS has upgraded its Elastic MapReduce offering to Hadoop 2.2 with the YARN framework. More updates and upgrades with better performance will follow. Architects should keep themselves abreast of the latest developments so that they can recommend the right products to their customers.

APPENDIX A

References and Bibliography

1. DataWarehouseBigDataAnalyticsKimball.pdf:
    http://www.montage.co.nz/assets/Brochures/DataWarehouseBigDataAnalyticsKimball.pdf

2. Big Data Diversity Meets EDW Consistency for New Synergies in BI: Nancy McQuillen,
    2 December 2011, www.gartner.com/id=1865415

3. http://blogs.informatica.com/perspectives/2010/11/17/understanding-data-integration-patterns-simple-to-complex: David Linthicum, November 17, 2010

4. http://www.datasciencecentral.com/profiles/blogs/11-core-big-data-workload-design-patterns: Derrick Jose, August 13, 2012

5. http://highlyscalable.wordpress.com/2012/03/01/nosql-data-modeling-techniques:
    Ilya Katsov, March 1, 2012

6. Big Data Analytics Using Splunk: Peter Zadrozny and Raghu Kodali (Apress 2013)

7. Securing Big Data: Security Recommendations for Hadoop and NoSQL Environments:
    Securosis LLC

8. Big Data Analytics: What It Means to the Audit Community: Markus Hardy

9. Getting Started with Storm: Jonathan Leibiusky et al: O'Reilly Media

10. Hadoop Operations: Eric Sammer: O'Reilly Media

11. Hadoop the Definitive Guide: Tom White: O'Reilly Media

12. Hbase in Action: Nick Dimiduk: Manning

13. MapReduce Design Patterns: Donald Miner et al: O'Reilly Media

14. Hbase the Definitive Guide: Lars George: O'Reilly Media

15. Embedding a Database for High Performance Reporting and Analytics: Bloor Research

16. Big Data: Hadoop, Business Analytics and Beyond: Jeff Kelly, Nov 08, 2012:
    http://wikibon.org/wiki/v/Big_Data:_Hadoop,_Business_Analytics_and_Beyond

17. CAP Twelve Years Later: How the "Rules" Have Changed: Eric Brewer, May 30, 2012:
    http://www.infoq.com/articles/cap-twelve-years-later-how-the-rules-have-changed

18. Key Value Database: http://bigdatanerd.wordpress.com

19. www.practicalanalytics.wordpress.com

20. www.baselinemag.com

21. A Perspective on Database: Where We Came From and Where We're Going: The Bloor Group, http://www.databaserevolution.com/research/

22. Big Data Analytics Architecture: By Neil Raden

23. http://media.smashingmagazine.com/wp-content/uploads/2011/10/Plane-newest.gif

24. http://www.submitinfographics.com/full-size-infographics/image-153.jpg

25. NoSQL Databases: www.newtech.about.com

26. Is Data Modeling Relevant in a NoSQL Environment?: Robinson Ryan

27. MapReduce Patterns, Algorithms, and Use Cases: Highly Scalable Blog:- Ilya Katsov, February 1, 2012

28. Big Data: Hadoop, Business Analytics and Beyond: A Big Data Manifesto from the Wikibon Community: Jeff Kelly

Index

S

T, U

V

W, X, Y

Z

Get the eBook for only $10!

Now you can take the weightless companion with you anywhere, anytime. Your purchase of this book entitles you to 3 electronic versions for only $10.

This Apress title will prove so indispensible that you'll want to carry it with you everywhere, which is why we are offering the eBook in 3 formats for only $10 if you have already purchased the print book.

Convenient and fully searchable, the PDF version enables you to easily find and copy code—or perform examples by quickly toggling between instructions and applications. The MOBI format is ideal for your Kindle, while the ePUB can be utilized on a variety of mobile devices.

Go to www.apress.com/promo/tendollars to purchase your companion eBook.

Apress®
THE EXPERT'S VOICE™